The Magic Bands

A Guide to 6m and 4m Amateur Radio

by
Don Field, G3XTT

Radio Society of Great Britain

Published by the Radio Society of Great Britain, 3 Abbey Court, Fraser Road, Priory Business Park, Bedford MK44 3WH. Tel: 01234 832700. Web: www.rsgb.org

Published 2020.

ISBN: 9781 9101 9397 6

Editing and design: Steve Telenius-Lowe, PJ4DX

Cover design: Kevin Williams, M6CYB

Production: Mark Allgar, M1MPA

Printed in Great Britain by Hobbs the Printers Ltd. of Totton, Hampshire

Any updates or errata to this book can be found at www.rsgb.org/booksextra

Contents

Preface

IT'S HARD TO BELIEVE that the original *Six Metre Handbook* appeared some 12 years ago. Five years and several reprints later, it was time to update it, the most significant change being the addition of significantly more information about the 4m band, which had become accessible to many more countries. This is a trend that has continued and we are now seeing Middle Eastern countries showing up on 4m too.

Time moves on and by far the biggest change in recent years has been the explosive growth of WSJT, in particular the adoption of FT8 as the *de facto* way of working DX (and even day-to-day QSOs) on both bands.

There are other changes too, in the more widespread adoption of remote working, in the software and internet resources available and, of great interest, our understanding of propagation on the band, largely as a result of FT8 highlighting more marginal band openings than previously. Not to mention the arrival of some excellent new radios with 6m and, increasingly, 4m capability.

And, courtesy of Justin Johnson, G0KSC, we have two 5-element OP-DES Yagi designs developed especially for this book.

Therefore, while by no means a new book, with plenty of good material carried over from the earlier editions, there is lots of new content to reflect what has been happening on the band in recent years. Which, of course, goes to show that 6m is constantly evolving as technology changes and, indeed, as a new generation of operators discovers the band.

My first exposure to 6m was on a visit to the MIT club station, W1MX, back in 1972. There was a 6m contest running and plenty of activity. Back then we didn't have access to 6m in the UK, for reasons which are explained elsewhere in this book. Eventually UK stations were allowed to apply for special permits for the band (a limited number of such permits were available) and then full access came in 1986, albeit with some power and antenna restrictions.

My own first foray on to the 6m band was in March 1988, with an elderly Yaesu FT-726R transceiver (about 20 watts) and a rather tired 5-element Tonna antenna which I had picked up cheap at a rally. My early QSOs were around the UK but then, in June, came my first Sporadic E QSO, with ZB2IQ, quickly followed by 9H, CT, PA, DL and LA. On 6 June 1988 I made my first QSO 'across the Pond': W1WHL at 2243UTC, followed by eight others that same night (including VE1YX and the late W3ZZ, two of the stalwarts of

50MHz). These late night (European-time) transatlantic openings in June and July are a regular feature of 6m and one of the joys of operating the band – they seem to be just as regular at sunspot minimum as at sunspot peak.

The excitement was mounting at that time, with stations in countries awaiting 6m permits getting their share of the action by working crossband. In June of 1988 I see that I worked DL, YO and 4X, for example, all of whom were on 28MHz and responding to my 6m CQs. Most of the 10m activity was around 28885kHz, which had become the meeting point for 6m enthusiasts, the spot where information was shared and DX stations showed up while waiting for the MUF to rise. The demise of 28885kHz for this purpose came some years later when the PacketCluster system effectively made it redundant, though there probably isn't the camaraderie on the Cluster system (and on chat rooms such as the ON4KST facility) that seemed to exist on '885.

Back in those halcyon days there was plenty to be worked, even with a modest station – I see in January 1989 I was again working across the Pond and in February 1989 worked ZS and J52US (Dave Heil, K8MN) as well as more US stations. And so it went on. 6m enthusiasm was growing around the world as more countries started to gain access. US operators were enjoying the opportunity to work DX outside the US, and there was great excitement among the 6m fraternity each time a new one appeared. The telephone lines were kept busy, with serious competition to be among the first to achieve DXCC on the band. For those of us in northern Europe, life was easier than nowadays, because there wasn't so much competition from better-located stations in southern Europe (neither Spain nor Italy had 6m access at that stage), although arguably the advent of FT8 has changed that because it's not necessarily the loudest station who gets through first. There are those who would also argue that the Cycle 22 peak was better for DX working than Cycle 23. Certainly, by that next peak I was finding life harder, possibly because propagation wasn't as good but also due to there being many more callers whenever a rare one appeared – so it was time to upgrade my station with a linear amplifier and a better antenna! And as for Cycle 24, we have seen no F2 propagation at all, at least from northern Europe.

During those years Six enthusiasts developed a strong sense of being part of something special. They met up at social functions and swapped war stories (often about the 'one that got away' during that fleeting opening while they were out doing the weekly shop!) And when they did meet they all recognised each others' calls because it was always the same folk in the pile-up and, often, the same guy they had worked at the distant end. Callsigns like CO2KK, HH7PV or HC5K were familiar to 6m operators as the only stations active on the band from their respective countries. There was a real challenge in working some of those new ones, too. Countries like HB9, initially, had access only in the wee small hours, when there was no

TV. QSOs often had to be made by meteor-scatter as there was little or no likelihood of Sporadic E at that time of night.

Things have moved on since those days, though readers who were there and have the tee-shirt will, I'm sure, recognise the atmosphere that I have described and perhaps be taken to reminisce about the 'good old days'. But 6m is still a challenge and a lot of fun as this book sets out to show. Nowadays there are new challenges, particularly with the widespread adoption of FT8. There is also more activity, with most countries now having a 6m allocation and most HF transceivers having 6m in their coverage. Some HF antennas even include 6m, the popular SteppIR range of HF Yagis, for example, having 6m as an add-on option. In response, most major expeditions now cater for 6m, especially if they are scheduled at a time of the year when Sporadic E or TEP propagation is likely. More and more are even catering for 6m EME. And the inclusion some years ago now of 6m in the ARRL DXCC Challenge listings has encouraged the more competitive HF DX chasers on to the band. So, all in all, a virtuous circle which makes for more activity and gives us more insights into 6m propagation as the years go by.

As for the 4m band, my experience goes back many years earlier than 6m, simply because 4m was very much part of the UK scene when I was first licensed in 1968, long before 6m became available to us. It was a band that I always enjoyed as part of the RSGB's VHF Field Day events and, from home, I was for a time the only 4m station active from Northamptonshire, so was popular with those who chased counties on the band. When 6m came along on a permanent basis I rather abandoned 4m although I used it for PacketCluster access for a while in the 1990s and, more recently, was able to participate in my local radio club net on 4m FM by way of a converted PMR set. But in the past couple of years I have returned to DXing and contesting on the band with a Yaesu FT-847 and, later, an Icom IC-7300 and small Yagi and have enjoyed making plenty of QSOs around Europe during the Sporadic E season. Like many of you, no doubt, I aim to build on that in the years to come.

There are some who worry that 4m will become just like any other band, losing its unique identity. Certainly, it has always enjoyed something of a special status with many regular users getting to know each other as friends. And, because there was relatively little commercial equipment available, it is a band that has encouraged homebrew construction. Balanced against this, though, there are new opportunities to explore propagation at that frequency and, in any case, who would want to deny those who are gaining the use of the band for the first time the enjoyment they are going to have. A complication remains, in that allocations vary significantly from country to country – full harmonisation is still some way off.

Amateur radio operators have traditionally grouped their frequency bands according to their characteristics for communication. The low frequency bands, good for night-time propagation but suffering from D-layer absorption during the day. The high-frequency bands, enjoying the benefits of daytime energisation of the ionosphere by the sun to support long-distance propagation. The VHF bands, essentially line-of-sight except under occasional situations of enhancement. The microwave bands, requiring specialist equipment and with yet another set of characteristics. Like all such groupings, life is never quite that simple. But 6m is almost unique in the way it behaves, sometimes acting as an HF band, with world-wide propagation, at other times acting much more as a VHF band, enjoying the benefits of Sporadic E, meteor scatter and other occasional propagation modes.

Because it has so many facets, 6m is both a challenge and an enigma and draws amateurs from both the VHF and HF worlds, especially those who want to try something new and different. 6m enthusiasts quickly become passionate about what they describe as 'The Magic Band' and it is hard to imagine any other waveband spawning its own club in the way that the UK Six Metre Group and SMIRK have arisen to cater for the interests of 6m operators.

The original *6 Metre Handbook* came about for exactly the same reason. Who would imagine a book specifically about 15m or even 2m? But 6m is sufficiently varied in what it has to offer, sufficiently unique in its characteristics, that it has a following all of its own. Increasingly, the same could be said of 4m, albeit for slightly different reasons, mainly that it is new to many and therefore an opportunity to learn, experiment and chase some new ones. It certainly enjoys the vagaries of Sporadic E propagation but we cannot expect F2 to rise this high in frequency, even if it eventually returns to 6m.

One of the difficulties in writing a book such as this is that the audience varies from those who have never been on Six or Four and want a fairly detailed guide, to those who have 200+ countries (on 6m, at least) and could well teach the author a thing or two. The former want some basic advice, the latter perhaps want to reminisce a little. Equally, some readers will have an extensive amateur radio library covering the more general issues of propagation, operating, QSLing and so on, whereas others may be looking to these pages to give them some basics. So, what I have tried to do is to start with the basics which, for some of you, may be old hat, so please be indulgent. For those who want to take these aspects further, there are plenty of useful references, many of which are listed in the bibliography, for example the ARRL and RSGB operating manuals, the ARRL and RSGB handbooks for more technical information, *The ARRL Antenna Book*, the *RSGB VHF / UHF Handbook* and a host of specialist propagation texts. Beyond the basics, this book covers facets of both bands such as history, a look at the 6m band from the perspective of some of the high achievers, use of weak-signal modes and much more. Hopefully something for everyone.

That said, while almost all the feedback I received on the earlier editions of this book was positive, which is heartening, the one minor criticism I got from a couple of people was that the original book didn't feature circuit diagrams and constructional details for 6m and 4m equipment. This was quite deliberate and I make no apology. I see this very much as a resource book to motivate, suggest ideas for improvement, fill in some historical background and so on. But constructional material is readily available elsewhere, in RSGB, ARRL and other publications as well as on the Internet. And, to be frank, and although we may wish it otherwise, probably 90% of those who turn to 6m and even 4m nowadays will be using ready-made equipment, their interest primarily being operating rather than construction. The one area that home construction remains popular is with antennas, which is why I am so pleased that we have a couple of new antenna designs for you to try.

I must thank those who have made the book possible. There is a long list of those who contributed to the first two editions and, in this revised book, I have also made a point of featuring some of the leading stations on 6m, with their stories of what they have achieved and how they have gone about it. My thanks to one and all.

I particularly, though, want to thank Tim Kirby, GW4VXE, who has proof-read several of the chapters for this one, pointing out the more serious errors and omissions! However, any remaining errors are definitely down to me. And I also want to thank Justin Johnson, G0KSC, for providing those two great Yagi designs.

I hope that the book will serve as a source of inspiration and enjoyment to many and particularly to those who are new to one or both bands. If so, I'm sure you will find operating on them a lot of fun.

Don Field, G3XTT,
September 2020

Chapter 1

Introduction

IT'S HARD TO BELIEVE that the UK only gained access to the 6m band in the 1980s and many other countries since. Nowadays, almost every country has access to the band and with that has come a huge increase in 6m activity along with recognition of the band for the ARRL DXCC Challenge, lots of interest in contests and, perhaps most significantly, a wholesale adoption of data modes.

Similar trends have occurred on the 4m band, albeit not to the same extent, partly because the more widespread availability of the band is more recent and, partly, I suspect, because it is not yet available in the USA.

However, what we have seen as a result of increased interest in these bands is their availability on many new transceivers. 6m is almost universal now on HF transceivers while a number also include 4m, at least in their European versions.

Although we are currently at the bottom of the sunspot cycle, the good news is that things can only improve over the next few years! That said, the last cycle did not give us the F2 propagation on 6m that we had seen in previous cycles. However, what we did see and continue to see is lots of DX being worked by extended Es propagation, particularly while taking advantage of the new digital modes. It seems that, at the right time and under the right circumstances, DX can be worked over significant distances as are achievable via F2 (although F2 appears to remain preferable, for example by way of the UK contacts to KH6). What we have also seen is an increase in EME (Earth-Moon-Earth, also referred to as 'moonbounce') activity on 6m, again facilitated by the digital modes. While 6m EME was, for many years, the preserve of those with a stack of at least four Yagis, results are being achieved nowadays by those with much simpler setups. And the good thing about EME, of course, is that it isn't confined to particular months of the year. Even if you don't work EME, meteor scatter propagation can also offer long-distance contacts year round, again helped by the newer data modes.

However, I don't want readers to get into a way of thinking that either 6m or 4m is only for DX working. Both are excellent bands for day-to-day 'ragchewing', EMERCOMMs and much else.

In these pages I aim to introduce both bands, building on the previous two versions of this book, the first of which appeared back in 2008. A lot has happened since then, in terms of equipment, operating, antennas and achievements.

I hope here to give you a flavour of how we come to be where we are today, how to set up a station and how to operate on 4m and 6m. And while I can't expect to cover everything you may want to know, I can hopefully point you in the right direction, to the wealth of information available on the Internet in particular.

THE MAGIC OF 6M

Not for nothing is 6m often referred to as the 'Magic Band'. If you are a TV broadcaster, think what a nightmare it is to be using part of the spectrum where propagation is line of sight for much of the time but then, especially at certain times of the year, the signals travel much farther, resulting in mutual interference between stations that normally are well clear of one another. Those of us of a certain age will recall the days of VHF television when, especially in the summer months, our regular TV station suffered severe interference from stations in adjacent states or countries, often described to the uninitiated as 'continental interference'. What a disaster!

But for amateur radio operators, this sort of variability is exactly what makes the hobby so interesting. You might have been listening to a quiet band on and off for weeks, maybe just holding regular schedules with a few friends in the neighbourhood. Then you turn on one day and the band is full of signals, many of the stations speaking in other languages or with strange

JW ON SIX!

(This short piece, by Chris Gare, G3WOS, gives a flavour of the excitement when a 'new one' shows – most 6m enthusiasts will have similar stories of their own.)

ON THE EVENING of 10 July 1996 at around 2030UTC the JW beacon started coming into the UK at some considerable strength. For over half an hour it varied between S4 and S7. Neil, G0JHC, rang Matt, JW5NM, only to discover that he was not at home. Later Chris, G3WOS, in desperation rang Ole, JW8GV, who then said he would go to the shack. He said he was not too optimistic about getting the station on the air for some reason. Neil explained to me later that they needed to disconnect the beacon antenna and connect it to an HF transceiver / transverter arrangement. After ten minutes, which seemed to be an eternity, the beacon stopped keying for a few seconds and then restarted. Are they in the shack? A couple of minutes later the beacon shuts off! They are there! A quick call to Neil revealed that they had already faded out in the north of England, much to his annoyance.

Tuning the band suddenly reveals Peter, JW7QIA, calling CQ on 50.090MHz. I called but there was no response – wasn't he hearing me? This happened about six times, then suddenly I get a "G3W?". It was then but a few minutes to complete a QSO and get a "TU". Mayhem then broke out and at least G3FPQ and G3IBI got into his log. The QSOs took place at around 2120UTC. As it turns out, Peter was only visiting the island and working a few stations while there.

accents, using prefixes that you normally only hear on the HF bands. And yet, even in the quiet times, you are aware that some of the more serious enthusiasts are still managing to make long distance contacts, bouncing their signals off meteor trails or even the moon. And rarely now on the traditional modes of SSB and CW, but using the latest signal processing software on their PCs to decode data signals quite inaudible to the human ear. Indeed, FT8 in particular has come of age in the 2019 and 2020 seasons.

For many years, the American Radio Relay League (ARRL) held off introducing a version of the DX Century Club certificate for 6m, thinking that it would be demotivating to have an award that was impossible to achieve. Hard to believe this is the same band where quite a few serious DXers now have country scores of well over 200 and working more than 100 countries in a year is achievable.

Six-metre and, increasingly, four-metre enthusiasts are a special breed. Not for them the relative predictability of 20m. They are prepared to listen to white noise for hours on end in the hope of maybe a 30-second window of opportunity where a DX signal rises above the noise just long enough to allow a contact to be made. The resulting sense of achievement really sets the adrenaline flowing, and is undoubtedly what makes these bands so popular. Clubs such as SMIRK (Six Meter International Radio Klub) and the UK Six Metre Group have much larger membership numbers than many similar organisations, reflecting the camaraderie between those who enjoy the special appeal of Six. In some ways the appeal is similar to that of 160m (Topband), where dedication, combined with an above-average level of technical knowledge and propagation expertise are required to reach the higher echelons of achievement. Those who populate the band are simply not satisfied with chatting through their local repeater on 2m or having reliable long-distance propagation on the major HF bands. They want a real challenge. To get a flavour of the intensity with which 6m enthusiasts follow their particular aspect of the hobby, take a look at some of their dedicated websites, good examples being those of M5BXB, G3WOS, JA1RJU and the like. Some, like VK3SIX and K1SIX have even selected callsigns that advertise their life-long love of 6m.

SOME HISTORY

Before 6m became available to the amateur radio community, amateurs in some countries were able to use the 56MHz (5m) band, a convenient frequency harmonically related to the lower bands (3.5, 7, 14, 28, 56MHz – no 21MHz back then). In the USA the 5m band became available in 1924, along with the 80, 40 and 20m bands (10m was authorised in 1927). The October 1924 issue of *QST* featured a 5-metre oscillator design, probably the first VHF constructional article to appear in print in an amateur radio publication. In the June 1929 issue of the UK magazine *Wireless World*, G6TW was reported to have won a prize, offered in the April issue of the magazine, for the first communication over 10 miles on 5 metres (presumably the first in the UK, given that 5 metres was already five years old in the USA).

The 13 December 1935 issue of *Wireless World* carried an article by a D R Parsons entitled 'Range of Five Metre Transmissions', in which he postulates ways in which 5-metre signals might propagate beyond the 'theoretical' limit (at that time it was thought that the ionosphere was unable to support reflections at such 'ultra-short' wavelengths). But it was in just the following year, 1936, that G5BY was the first European station to span the Atlantic on 56MHz, when his signals were heard by W2HXD, although no two-way contact ensued.

By 1938 the 56MHz distance record was held by W1EYM and W6DNS, a 2500-mile path. With the advent of war in 1939 activity on all bands was much reduced, with many countries closing down amateur radio activities completely. Technological developments during the war years had made it cost-effective to roll out TV broadcasting on VHF frequencies that, prior to the war, were considered to be of little use. Thus, immediately post-war there was quite an administrative battle over the use of the VHF spectrum.

Interestingly, the then ARRL General Manager, Kenneth Warner, during his 1944 negotiations with the FCC, is quoted as having said, "We have previously characterised the performance of this band to you as being erratic, unpredictable, unreliable and unexpected, a band where anything can and generally does happen, and we have explained that its very eccentricities give it a peculiar charm for us, though they make it singularly bad for regular service."

In the USA the outcome of the negotiations was that amateurs lost the 56MHz band, which now fell within the 54 – 60MHz spread of TV's channel 2 but managed, instead, to get an allocation between 50 and 54MHz (TV channel 1 was 44 to 50MHz). The new 6m band, as it was now known, became available in the USA in March 1946 and activity soon picked up, with the first transcontinental QSO taking place on 14 June, between W6OVK (CA) and W2BYM (NJ).

Before the year was out 6m was available to amateurs in Canada, Australia and New Zealand. By 1947 there was further activity, especially from Central and South America and some limited authorisations became available in Europe (the Netherlands, with 100 watts, and the UK with 25 watts. UK operators were limited to certain times of the day to avoid

UK FIVE-METRE FIRSTS
(source: *Short Wave Magazine*, May 1947)

France	G2FA / F8NW	29 March 1936
Italy	G5MQ / I1IRA	2 July 1938
Holland	G2AO / PA0PN	17 August 1939
North Africa	G5BY / FA8B	24 June 1946
Switzerland	G5BY / HB9CD	22 August 1946
Sweden	G5TH / SM5FS	24 May 1947
Denmark	GM6KH / OZ7G	24 May 1947
Belgium	G6DH / ON4KN	25 May 1947
Czechoslovakia	G5BY / OK2MV	22 June 1947
Malta	G6LK / ZB1AB	30 June 1947
Gibraltar	G2XC / ZB2A	22 July 1947

interference to TV services and the authorisations were available until 30 April 1948, see below). Immediately post-war the UK retained a 5m allocation, at 58MHz, and a number of tests were conducted with the hope of achieving a cross-band 5m / 6m QSO. This didn't happen and the first trans-Atlantic QSOs were made cross-band 10m / 6m, simply because while these tests were taking place the MUF never reached as high as 58MHz. Hence a request was made by UK amateurs to the UK General Post Office for a temporary 6m permit. As it happens, the 5m band was only available in the UK for about three years before those frequencies, as in the US, were required by the emerging national TV broadcasting service. Things didn't look good. In an editorial in the May 1947 issue of *Short Wave Magazine*, Austin Forsyth, G6FO, looking ahead to the Atlantic City conference, at which frequency allocations were to be discussed and decided, wrote, "To 5-metre operators, the lack of any suggestion between 29.7 and 168 mc is disastrous. A band, however narrow (250 kc would do), in the 50 or 60 mc region, allocated on an international basis, would be sufficient for our needs". It didn't happen. Amateurs did gain the 2m band, but anything between 29.7 and 144MHz was left to national licensing authorities, with the result that, for something like the next 40 years, there was no internationally-harmonised band.

Nevertheless, UK amateurs continued to lobby for a lower VHF band to replace 5m and in 1956, seven years after 5m was taken away, were allocated 70.2 to 70.4MHz (or Mc/s as it would have been referred to at that time), the 4m band. Unfortunately this was available only in the UK and a limited number of UK overseas territories, so British amateurs continued to press for a 50MHz allocation but, as with much of the rest of Europe, this didn't materialise (other than by way of occasional special permits) until the TV broadcasters finally started to abandon VHF in favour of their newer and more effective UHF allocations (more effective because they offered enough bandwidth for higher definition and colour, but also because they got away from the vagaries of VHF propagation – all that dreadful Sporadic E and stuff, which was exactly what the amateurs were interested in).

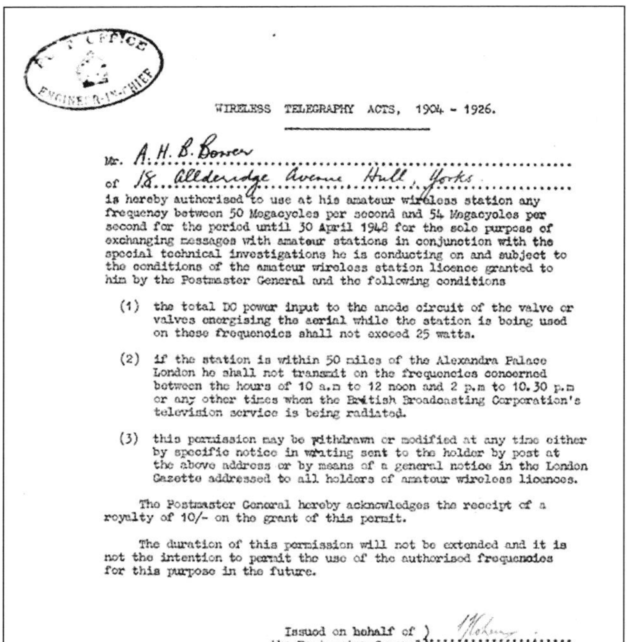

POST OFFICE ENGINEER-IN-CHIEF

WIRELESS TELEGRAPHY ACTS, 1904 - 1926.

Mr. A. H. B. Bower

of 18 Allderidge Avenue Hull, Yorks

is hereby authorised to use at his amateur wireless station any frequency between 50 Megacycles per second and 54 Megacycles per second for the period until 30 April 1948 for the sole purpose of exchanging messages with amateur stations in conjunction with the special technical investigations he is conducting on and subject to the conditions of the amateur wireless station licence granted to him by the Postmaster General and the following conditions

(1) the total DC power input to the anode circuit of the valve or valves energising the aerial while the station is being used on these frequencies shall not exceed 25 watts.

(2) if the station is within 50 miles of the Alexandra Palace London he shall not transmit on the frequencies concerned between the hours of 10 a.m to 12 noon and 2 p.m to 10.30 p.m or any other times when the British Broadcasting Corporation's television service is being radiated.

(3) this permission may be withdrawn or modified at any time either by specific notice in writing sent to the holder by post at the above address or by means of a general notice in the London Gazette addressed to all holders of amateur wireless licences.

The Postmaster General hereby acknowledges the receipt of a royalty of 10/- on the grant of this permit.

The duration of this permission will not be extended and it is not the intention to permit the use of the authorised frequencies for this purpose in the future.

Issued on behalf of the Postmaster General

Date 29 JAN 1948 1948.

The late Brian Bower's, G3COJ, 50Mc/s permit, dated January 1948.

Anyway, back to the 6m story. In the immediate post-war years J9AAK was active from Okinawa, and amateurs in other parts of the world without 6m privileges were busy looking for cross-band contacts (10m to 6m). The first trans-Atlantic QSOs were in spring 1947 and some other rare DX was worked from Europe including MD5KW (Suez, operated by G5KW) and SU1HF (Egypt). The following news prompted a separate 'news flash' to be included in the April 1947 issue of *Short Wave Magazine*, the only occasion I have seen this unusual step taken, so the news was clearly considered highly memorable, "At 1330 on March 26, PA0UN was received on 50 mc by ZS1AX, ZS1P and ZS1T, at up to S9". At this time activity was confined to AM and CW. SSB didn't really make inroads on to 6m until the mid-60s or so. Activity continued to increase and, in 1956, the distance record was extended to 12,000 miles with a QSO between LU9MA and JA6FR. Give that this is half the circumference of the world, it was already clear that 6m supported world-wide propagation when the sunspot cycle was co-operating (that cycle is remembered by those who were around at the time as being quite exceptional, as indeed it was). The main limitation to a high country score on 6m was lack of 'new ones' to work, with most of Europe, for example, being off-limits because that part of the spectrum was used for TV broadcast purposes. That said, some special authorisations came into being in 1957 on the back of International Geophysical Year, bringing activity to the band from the Azores, Madeira, Portugal, Norway, Sweden and Poland.

It was not until the 1980s that things started to change in Europe. In the UK a proposal emerged in 1980 to allow a limited number of amateurs access to the band. This move was supported by the Home Office (the licensing authority at that time) but an objection was received from the BBC and the plan was abandoned. Two years later things had moved on. The UK Six Metre Group (UKSMG) had been founded by Steve Richardson, G4JCC, to promote interest in the band (see sidebar on page 16) and Jim Sleight, G3OJI, a senior engineer at the BBC, helped to overcome objections from that organisation. Some 300 UK amateurs applied for experimental licences for the band, to be whittled down to 46, the number that the Home Office had agreed to issue. In December of 1982 the GB3SIX beacon was licensed for 24-hour operation and in January 1983 the first permits were issued, allowing operation from 1 February outside TV hours (those were the days when TV actually closed down overnight – hard to believe now). By the end of the year this restriction had been lifted and 24-hour operation was allowed and in February 1984 it was announced that a further 60 permits were to be issued, bringing the total to 100.

Lots of activity ensued, despite limitations on ERP and antenna height, with plenty of the permit holders managing to work 'across the pond'. This occasionally led to some interesting situations, such as with my good friend Dennis Robinson, GJ3YHU, calling a W5 station on the band who refused to work him on the basis that GJ3YHU wasn't a 'proper' callsign (the W5 presumably never previously having heard any stations from outside the

THE CREATION OF UKSMG

(Abridged from a 2008 letter by Steve, G4JCC, after he had been made an Honorary Member of UKSMG in recognition of his efforts in founding the Group and contributing to its early success).

THE AUTUMN of 1979 produced some outstanding propagation on six metres and I hold a tape containing a long QSO by Bob, VE1YX, with another Canadian operator on Six, recorded at that time, together with some other QSOs. These openings generated a lot of interest in VHF propagation and I had many conversations with Harold Rose, G8NWF, soon to become G4JLH. One of the points that arose was that there was great interest in six metres during the solar maxima but interest waned and people went to other bands and pursuits as propagation declined. Around this time I learned, before it was public knowledge, that all low-band television was to be switched off. I took this up and my contact promised to keep me informed of developments. Later he told me that the BBC wished to retain all their TV frequencies which meant the whole band from *circa* 40MHz to 80MHz. This information was very confidential at the time so I could make no direct use of it without compromising my informant, who was not a ham nor was he interested in amateur radio. However, it was obvious that there was an opportunity to regain a VHF band (now 6m) which we had lost to the BBC (this was the old 56Mc/s band; I had been a SWL on this frequency prior to WWII). So I decided to contact and have some discussions with amateurs who were or could be interested in a VHF band at 50MHz. I talked to G4JLH and, most importantly, G5KW with whom I had a very useful and productive meeting.

My discussions with Ken, G5KW, centred round the best course to pursue to retain the interest of radio amateurs. We were agreed that a few temporary licences and a beacon or two were not sufficient to retain this interest. Ken suggested that a column in *RadCom* about 6m matters might be the way to go or to form a branch of the 'Six Klub' which was an American organisation based in South Texas. So gradually our thoughts turned to having our own 6m club. I agreed to publicise this idea and gradually it grew but it needed a title, anything containing the word "Club" was out as it could be confused with the US one, so gradually the title UK Six Metre Group was adopted; then I began to write letters and solicit members from 6m operators (cross-band of course). The rest is history but it took a lot of discussion and work to reach this stage. The Merriman Committee was formed by government to decide what use should be made of the vacant TV frequencies and I, along with several other members, made representations to the Committee, which resulted in a recommendation for a band at 50MHz.

US). From 1 February 1986 all UK Class A licensees were allowed access to 6m, albeit with an ERP limit of 100 watts on SSB (25W on CW) and a maximum antenna height of 20m, horizontally polarised only. The main fear was that UK amateurs using 6m could cause interference to TV in mainland

Europe, particularly in Belgium and France. In reality no problems seemed to arise and Class B licensees were given access to the band from 1 June 1987. From then on the various limitations were incrementally relaxed as VHF TV was gradually closed down not only in the UK but across Europe.

Similar stories were being enacted in many other countries throughout this period, usually with some sort of special permit in the first instance, allowing the national licensing authority to withdraw the band should problems occur. Once it was clear that all was well restrictions would gradually be withdrawn. Nevertheless it took something like 25 years before all of Europe, for example, had 6m access (Hungary being the last major country to gain access to the band, in 2005, though some, particularly European Russia, are currently inactive once again).

For a much more detailed account of 6m history, by far the best starting point is the series of articles written by Ken Willis, the then G8VR, for *Six News* (the UKSMG journal) and reproduced on the UKSMG website.

WHAT DOES SIX HAVE TO OFFER?

Six metres demonstrates the characteristics of both a VHF band and an HF band at different times. For much of the time the MUF is below 50MHz and propagation on Six is predominantly line-of-sight, as on the higher VHF bands. At such times the band can be handy for local 'ragchews', particularly if you don't have other VHF equipment in the shack but have an HF transceiver with 6m capability (which most do nowadays). There is no simple answer as to what you might expect the range to be, as it depends very much on your location and the type and height of antenna that you are using. But with a decent home-station setup you might expect to be able to work similarly-equipped stations up to 100 miles (160km) or so without too much difficulty on CW / SSB and somewhat further using data modes.

In many countries there is a network of FM repeaters to extend the workable range, exactly as on the other VHF bands. But don't expect much mobile activity on 6m; mobile antennas are much more manageable on 4m, 2m and 70cm (and 220MHz in North America), although a good 6m repeater will typically provide more extensive coverage than those on 2m and 70cm. Some of these repeaters are connected to the Internet via systems such as *Echolink*, giving users access to the whole world via their 6m access channel. Up-to-date lists of repeaters can be found on the Internet. A list of UK 6m repeaters also appears at Appendix C to this book, current at the time of writing.

A number of channels is also allocated to packet radio, for applications such as messaging, PacketCluster, APRS and Satgate. It is not the intention of this book to cover these specialist applications, which are well documented elsewhere and not specific to 6m, but the message is that 6m carries much the same sort of day-to-day activity as other VHF bands, applications which generally do not exist on the HF bands, largely due to bandwidth limitations.

But even when there is no 'skip' on 6m, whether from F layer, Sporadic E or other ionsospheric effects, you will hear stations working long distances by a variety of means. These are principally EME ('moonbounce') and meteor scatter. The names are self-explanatory and these propagation mechanisms are covered more fully elsewhere in this book. They help to keep alive interest in long-distance working on Six, even when the band is otherwise flat. EME requires a well-equipped station but meteor scatter is possible on almost a daily basis, even with a modestly-equipped station, particularly if you are using the appropriate WSJT software (again, covered later in this book). During the main meteor showers this mode becomes even more effective, permitting contacts up to something like 1400 miles (2300km).

For many 6m operators, though, the exciting times of the year, particularly during the low phase of the sunspot cycle, are those when Sporadic E (Es, or 'E skip') propagation occurs. The mechanism is explained in Chapter 4 although even now there is some argument about exactly what causes Sporadic E. But from an operational point of view the key aspect is that Sporadic E occurs annually at very predicable times of the year. In the northern hemisphere the main Es season is in the May to August period but certain openings seem to be focused on an even more clearly-defined window in late June to early July, when, for example multi-hop and extended Es support trans-Atlantic, coast-to-coast US and occasionally trans-Pacific contacts. As a result, many 6m DXpeditions are scheduled for this time of the year and serious 6m enthusiasts plan to take vacation time so that they can be on the band when openings occur. Signal strengths during Sporadic E openings can be high, so that even the most modest station can make successful contacts during these openings.

For operators in the lower latitudes (closer to the equator), for example in Southern Europe, trans-equatorial propagation (TEP) is also a reliable means of long-distance propagation for much of the year and this is covered elsewhere in this book.

Somewhat less predictable is auroral and auroral E propagation, again described in the chapter on propagation. Auroras can occur at any time of year but they tend to be more common around the vernal and autumnal equinoxes. They follow major solar disturbances, so the keen 6m DXer will keep an eye not only on the bands (a strong aurora leads to a blackout of the HF bands) but also on the various space weather sites on the Internet.

The previous paragraphs have described how 6m can bring some interesting contacts year in and year out. But the fun really starts during years of high solar activity when the F2 MUF reaches 50MHz and 6m starts to behave like an HF band, with F layer propagation. When this happens world-wide contacts are possible. Sadly, Cycle 24 did not produce the sort of MUFs that were experienced during the previous solar maximum, and it is easy to forget how Six can come to life at the peak of the cycle, as it did back around the turn of the millennium. At that time European 6m operators were working regularly into the Far East and Australia, North American amateurs were seeing regular openings across the Atlantic, out into the Pacific and across their own continent. Occasionally some really long- haul contacts were made, such as the long-path contact between the late David Courtier-Dutton, G3FPQ, and KH6SX in Hawaii, at 0957UTC on 7 April 2002, a path of about 26,500km (and only worked once previously from the UK, between KH7R and G4EAT on 27 March 2000, a great opening in which many other European stations worked into KH6). These band openings can be fleeting and propagation will tend to peak at certain times of the year. In a sunspot maximum year F2 propagation begins in October, peaks in the month prior to the winter solstice and dies out sometime near the end of January. It is this type of propagation that allows high country totals to be achieved on 6m as it is only F-layer propagation that will allow truly long-haul contacts on 6m, at least unless you are equipped for EME (but far fewer countries are likely to be activated on EME than will be active during F-layer openings). That said, the advent of FT8 has made possible some remarkable QSOs on the band that simply would not have been feasible on other modes. In the 2020 season, for example, UK stations have been regularly working into JA, VR2, BV, HL and VK and, to the west, into W6, W7 and VE7.

SOME 6M ACHIEVEMENTS

To put some of the foregoing in perspective, it is worth a short diversion to review some of the operating achievements on 6m to date. The first 6m DXCC awards were issued to K5FF and W5FF in January 1990. VE1YX was third and JA4MBM fourth to receive the award. The first European station to achieve DXCC on the band was G4AHN in 1991, receiving certificate number 29. VHF operators are also keen grid chasers and by way of example, K1TOL has over 1600 grids confirmed on the band.

When I prepared the last edition of this book (April 2013), I noted that there were some 48 amateurs with 200 or more entities credited on 6m (up from 22 in the previous edition of this book, despite the somewhat flat propagation in the intervening years). However, that figure is now an astonishing 91, quite remarkable given the lack of F2 propagation in the meantime. Not surprisingly, the listings are dominated by European stations, given that European operators have an 'easy' 70 or so countries on their doorstep. The highest placed JA is JA1BK with 245 entities credited while the highest North American is K2ZD with 213 credits. W7GJ, who rarely experiences Sporadic E propagation in his mid-western location, comes in at a very creditable 204, largely due to his focus on EME activity.

Peter Sprengel, PY5CC (now PP5XX), stands at 230, the highest-placed South American. It is a number of years since Peter was active on Six, but prior to that there was little if anything he had missed on the band. Peter's 200th country on Six was KH6ND/KH5 (Palmyra and Jarvis) in November 2000, making him almost certainly the first amateur to achieve 200 countries on the band. Peter lives in what must be the ideal location for 6m, positioned exactly right in respect of the geomagnetic equator to have reliable trans-equatorial propagation to most of the main areas of the globe and if he had been more active in recent years could possibly still be at the top of the list.

To put these scores in perspective, the UKSMG web page **[1]** shows 276 entities that have been worked from the UK (up from 250 in 2013). However, this total includes several deleted entities and three (1B, FR/G and YV0) that have not been accepted for DXCC. The total of available entities in recent years is, to a large extent, the result of EME activity from otherwise hard-to-work entities in the Pacific.

Most of these 200-plus scores were achieved with the help of a couple of sunspot peaks, but even at the bottom of the sunspot cycle it is astonishing what can be worked on the band. Even on CW and SSB, it is not unusual in the summer from the UK to hear and work US and Caribbean stations on 6m while there is nothing doing on the high HF bands. With FT8, much of Asia becomes workable too. 6m really is a unique band.

A QUICK TOUR OF 6M

The 6m band plan is covered in detail in Chapter 5. In broad terms, in Europe you will find most beacons near the bottom of the band (quite deliberately, so that as the MUF rises past 50MHz the beacons should be the first signals to be heard, though in the US the beacon band is 50.060 to 50.080MHz and there have been suggestions that this allocation should be harmonised globally, if only to clear the bottom end of the band for CW activity. Indeed, some beacons have now moved as high in the band as 50.400MHz). Then CW up to 50.100MHz and CW / SSB above that frequency. By convention, 50.110 is the acknowledged inter-region calling frequency (Intercontinental DX Centre of Activity). This is helpful when not much is happening on the band, as your receiver can be left on this frequency which should be the first port of call of anyone seeking contacts when there is the possibility of a band-opening. The idea is that it is an intercontinental calling frequency only and that once contact is established the stations should move to another part of the band. In practice this doesn't always happen and the frequency can quickly become chaotic. I have often heard stations in two or even three continents calling CQ or working stations simultaneously on 50.110, seemingly oblivious to one another and to the general confusion they are causing. Frankly, with Cluster available in most DXers' shacks and with many 6m DXers using transceivers with panoramic displays, it makes little sense to call on 50.110 once the band has opened. Call either side of that frequency (the 'DX Window' is considered to be 50.100 to 50.130MHz) and it won't be long before you are spotted by one means or another. And with the advent of FT8, 50.313MHz and 50.323MHz (the FT8 frequencies) can often be full of DX activity with little or nothing happening on SSB or CW.

Higher in the band are designated meeting points for cross-band, meteor scatter, digital modes, SSTV, and other specialist interests. Of particular interest nowadays are the centres of activity for machine generated modes

(MGM), particularly, as already mentioned, 50.313 and 50.323MHz (intended purely for intercontinental QSOs), where FT8 activity is concentrated. After that come designated channels for FM working, both simplex and repeaters, and priority channels for emergency communications. With 2MHz to play with in the UK (up to 4MHz in some countries, including the USA) there should be room for everyone, though the FT8 frequencies can fill quickly when there is a good band-opening. And it is important to be aware that some countries do not have the full band, while others are still limited to spot frequencies. To quote *Wikipedia*, "In some other countries, the band is restricted to military communications. Further, in some nations, the frequency range is used for television transmissions, although most countries have assigned those television channels to higher frequencies." As a consequence, although the International Telecommunication Union does not allocate 6-metre frequencies to amateurs in Europe, the decline of VHF television broadcasts and commercial pressure on the lower VHF spectrum has allowed most European countries to provide a 6-metre amateur allocation.

EARLY 6M EQUIPMENT

Emil Pocock, W3EP, contributed a fascinating article on the history of 6m equipment to *Six News* **[2]**, and the following draws heavily on that material. One of the earliest sets designed specifically for 6m was the Gonset Communicator or 'Gooney Box' as it was affectionately known. This set was launched in the '50s and used a dual-conversion superhet receiver, and crystal-controlled transmitter producing about 10 watts. The price in 1955 was $230. The Heathkit Sixer was also popular during the 60s, costing just $45 in kit form. It provided 5 watts of transmit power (AM and CW) but the receiver was a regenerative design, sensitive enough but not exactly suitable for DX working. There were quite a few other models from manufacturers such as Lafayette, Hallicrafters, Knight and Clegg. Not surprisingly these were all US or Japanese companies, those being the two major countries at that time to have access to Six.

The more serious 6m enthusiasts tended to build or buy a receive converter to work with their HF receiver and, for transmit, to build a separate, stand-alone transmitter unit, often crystal-

The 6m Gonset Communicator or 'Gooney Box'.

Copy of original advert for the Swan 250 SSB, CW and AM 6m transceiver.

controlled (those were the days when VHF operators would have just one or two crystals for transmit and, after calling CQ, would tune the band for possible replies).

Things improved enormously during the 60s when several US companies started manufacturing VFO-controlled SSB rigs with as much as 100 watts output. These included the Heath SB-110, the Swan 250 and, for those wanting the Rolls-Royce of 6m rigs, the Drake TR-6. This boasted 19 valves, 10 transistors and 12 diodes and used three 6JB6 valves in the final to put out a hefty 300 watts. But the price was $600 in 1969, not cheap.

As many readers will recall, the face of amateur radio equipment changed dramatically in the 70s with the first solid-state equipment, coming mainly from Japan, the US manufacturers being somewhat slower in adopting the new technology. In 1976 Yaesu introduced the FT-620B, the first mass-produced, solid-state 6m transceiver.

The TS-600 and IC-551 were equivalent models, launched by Kenwood and Icom respectively. One of the great benefits of solid-state equipment was that it made portable operation so much easier, with equipment easily powered from a car battery rather than requiring a generator. Another way on to 6m was with a transverter. Yaesu, for example, sold a transverter frame to work with the FT-101 / 901 series of HF transceivers. Power was drawn from the transceiver and the frame could hold one or more VHF / UHF transverters, one of which could be for 6m.

By the 1980s 6m-only equipment was in decline and it was becoming much more common to find 6m as a featured band in both multiband HF and VHF transceivers. But perhaps the greatest boon to 6m operators was when manufacturers started to include general-coverage receive capabilities, which allowed DXers to follow the MUF upwards, perhaps by monitoring commercial and TV signals in the 45 – 50MHz range. Nowadays almost all new transceivers include 6m as standard.

THE 4M BAND

When I put together the previous edition of this book, the 4m band was starting to become available in a handful of (mainly) European countries.

Indeed, even in the first edition (2008), I was able to report, for example, that the 1A0KM (Sovereign Military Order of Malta) station had been active during the 2007 Sporadic E season.

However, the band has developed significantly since then, with other countries freeing up spectrum and allowing their amateurs on to 4m, albeit sometimes with restricted frequency allocations or power. During the 2020 season, for example, UK stations were working into the Middle East (Israel, Bahrain, Kazakhstan) and down into Africa (the Canary Islands and Western Sahara).

The 4m band pre-dates 6m by some years as far as UK radio amateurs are concerned. As mentioned earlier, UK amateurs received an allocation at 70.2 to 70.4MHz back in 1956, by way of compensation for losing the 56MHz (later the 58.5 to 60MHz) band. Interestingly, a French website **[3]** records that, in 1948, 72 – 72.8 MHz was allocated to France (till 1961).

In the UK, the 4m band was later extended to 70.025 – 70.7MHz and subsequently moved to today's allocation of 70.0 – 70.5MHz.

The 4m Website shows a QSO between G5KW and G8KW, on 2 November 1956, as being a 'first'. 'Firsts' are also shown of QSOs with countries that include Algeria, Andorra and Ireland. Indeed, on the occasion of the International Geophysical Year 1957 – 1958, the following countries were allocated frequencies between 70 – 72.8MHz: Ireland 70.575 – 70.775MHz, Finland 70.2 – 70.3MHz, Germany 70.3 –

There have been occasional operations from SMOM, the Sovereign Military Order of Malta, as 1A0KM and 1A4A since 2007.

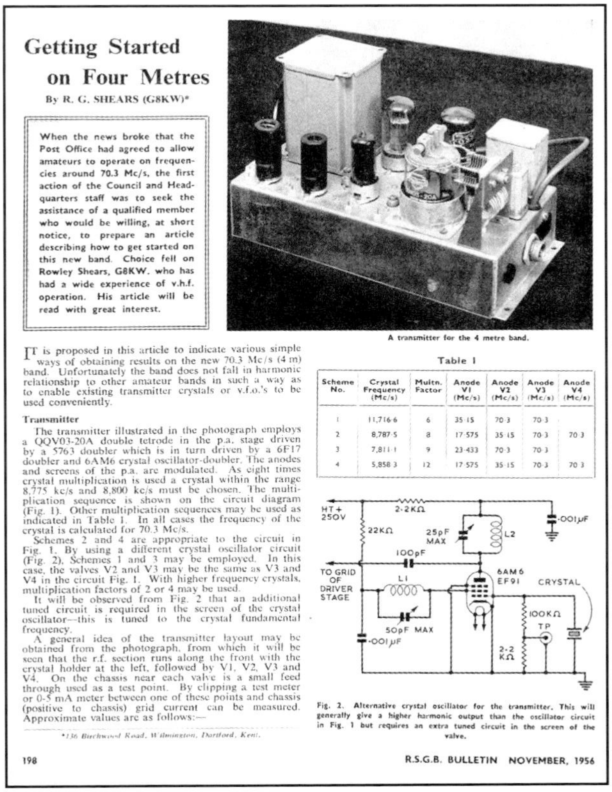

Getting Started on Four Metres

By R. G. SHEARS (G8KW)*

When the news broke that the Post Office had agreed to allow amateurs to operate on frequencies around 70.3 Mc/s, the first action of the Council and Headquarters staff was to seek the assistance of a qualified member who would be willing, at short notice, to prepare an article describing how to get started on this new band. Choice fell on Rowley Shears, G8KW, who has had a wide experience of v.h.f. operation. His article will be read with great interest.

IT is proposed in this article to indicate various simple ways of obtaining results on the new 70.3 Mc/s (4 m) band. Unfortunately the band does not fall in harmonic relationship to other amateur bands in such a way as to enable existing transmitter crystals or v.f.o.'s to be used conveniently.

Transmitter

The transmitter illustrated in the photograph employs a QQV03-20A double tetrode in the p.a. stage driven by a 5763 doubler which is in turn driven by a 6F17 doubler and 6AM6 crystal oscillator-doubler. The anodes and screens of the p.a. are modulated. As eight times crystal multiplication is used a crystal within the range 8,775 kc/s and 8,800 kc/s must be chosen. The multiplication sequence is shown on the circuit diagram (Fig. 1). Other multiplication sequences may be used as indicated in Table 1. In all cases the frequency of the crystal is calculated for 70.3 Mc/s.

Schemes 2 and 4 are appropriate to the circuit in Fig. 1. By using a different crystal oscillator circuit (Fig. 2), Schemes 1 and 3 may be employed. In this case, the valves V2 and V3 may be the same as V3 and V4 in the circuit Fig. 1. With higher frequency crystals, multiplication factors of 2 or 4 may be used.

It will be observed from Fig. 2 that an additional tuned circuit is required in the screen of the crystal oscillator—this is tuned to the crystal fundamental frequency.

A general idea of the transmitter layout may be obtained from the photograph, from which it will be seen that the r.f. section runs along the front with the crystal holder at the left, followed by V1, V2, V3 and V4. On the chassis near each valve is a small feed through used as a test point. By clipping a test meter or 0-5 mA meter between one of these points and chassis (positive to chassis) grid current can be measured. Approximate values are as follows:—

*136 Birchwood Road, Wilmington, Dartford, Kent.

A transmitter for the 4 metre band.

Table 1

Scheme No.	Crystal Frequency (Mc/s)	Multn. Factor	Anode V1 (Mc/s)	Anode V2 (Mc/s)	Anode V3 (Mc/s)	Anode V4 (Mc/s)
1	11,716·6	6	35·15	70·3	70·3	
2	8,787·5	8	17·575	35·15	70·3	70·3
3	7,811·1	9	23·433	70·3	70·3	
4	5,858·3	12	17·575	35·15	70·3	70·3

HT+ 250V
2·2KΩ
·001µF
22KΩ
25pF MAX
L2
100pF
TO GRID OF DRIVER STAGE
6AM6 EF91
L1
CRYSTAL
100KΩ
TP
50pF MAX
·001µF
2·2 KΩ

Fig. 2. Alternative crystal oscillator for the transmitter. This will generally give a higher harmonic output than the oscillator circuit in Fig. 1 but requires an extra tuned circuit in the screen of the valve.

198

R.S.G.B. BULLETIN NOVEMBER, 1956

When UK amateurs received access to the 4m band the RSGB asked VHF enthusiast Rowley Shears, G8KW, to write an article on how to get started on the band. His article, published in the November 1956 *R.S.G.B. Bulletin*, included circuits for a transmitter and receiver. Elsewhere in that issue there was also a converter for 4m.

MAGIC BAND OPENS WIDE TO ALLOW CROSS-BAND TRANS-ATLANTIC CONTACT

(from the weekly *ARRL Letter*, 29 June 2007)

ON MONDAY 25 JUNE [2007] there was a big opening to Europe on 6 metres (50MHz), also known as the 'magic band'. In most areas, the opening started in the local US mid-morning, lasting until dark. According to *QST* column author of 'The World Above 50MHz', Gene Zimmerman, W3ZZ, "Many areas of the country that do not normally work Europe, including the Midwestern states, west Texas and Colorado, worked stations in Europe. At one point near the end of the opening, stations on the East Coast of the US were working stations in Hawaii on the Big Island." In this opening, Mike Smith, VE9AA, in New Brunswick was on 50MHz when he worked Nigel Coleman, G7CNF, on CW cross-band; Coleman was on the 70MHz band (4 metres) in England. Zimmerman said, "Though a few cross-band contacts were made via F2 propagation during the sunspot maximum period in the 1970s, this is believed to be the first 50 / 70 trans-Atlantic cross-band contact ever made on multi-hop Sporadic E propagation." RSGB VHF / UHF Manager David Butler, G4ASR, concurred: "It certainly isn't the first United Kingdom – North America 4-metre / 6-metre cross-band QSO, but it probably is the first via multi-hop Sporadic E." In North America, 70MHz is channel 4 on television sets.

70.4MHz, the Netherlands 70.3 – 70.4MHz, Norway 70.6 – 72.0 MHz, Yugoslavia 72.0 – 72.8MHz and Austria 70 MHz special licences **[4]**.

In March 1993 the European Radiocommunications Office (now ECC) of the CEPT launched Phase II of a detailed spectrum investigation (DSI) covering the frequency range 29.7 – 960MHz. The results were presented in March 1995. Regarding the Amateur Radio Service, the DSI management team recommended (among other things) that 70MHz be considered as an amateur band.

For most of the intervening years, Four was available only in the UK and some countries with historic connections to the UK (Gibraltar, ZB2, being the obvious example). So the experience of 4m for many UK amateurs has been working around the UK (taking advantage, for example, of auroral propagation to work between stations in the south of England and stations in Scotland and Northern Ireland) and, in summer, enjoying Sporadic E contacts into those few Mediterranean countries with access to the band. To add a further level of interest, some 4m operators experimented with cross-band working to make contacts with other countries, which enabled data to be gathered about propagation on the band.

The late Jack Hum, G5UM, used his VHF column in the *R.S.G.B. Bulletin*

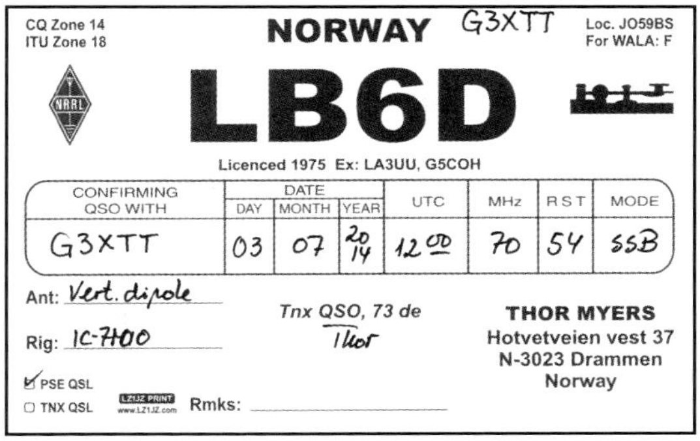

Many European countries now have access to the 4m band.

to encourage 4m activity although the band was never as widely used as might have been hoped (perhaps because, in many parts of the UK, there was still a clash with local TV frequencies). However, 4m activity during VHF Field Day in July was always of particular interest, with the hope that a ZB2 contact might be possible via Sporadic E.

In recent years, with the closing down of VHF TV across much of Europe, many European countries have now released allocations in the 70MHz part of the spectrum, albeit with allocations varying and, in some cases, not overlapping. Nevertheless, the majority of European countries and a handful elsewhere, particularly in the Middle East, now have access to the band. This has added greatly to the level of interest, spurred on by the increasing availability of equipment covering the band. And, of course, FT8 has added significantly to the activity levels on the band.

A QUICK TOUR OF 4M

The 4m bandplan is covered in Chapter 5 and is rather complex as national allocations vary so much. But the main centres of activity are 70.200MHz for SSB operation and, increasingly, 70.174MHz for FT8.

On a day-to-day basis there are also many FM nets on 4m, usually run by local radio clubs. My previous club, for example, has been able to source and convert Ascom and, later, Simoco PMR radios for use on 4m, which has led to an active weekly net taking place among club members. But where things are changing almost by the day is the level of DX activity, as more and more countries gain access to the band.

Propagation on 4m is in many ways similar to 6m, with the exception of F2 propagation, which is rare to non-existent and, to date, no reported EME activity. (However, it must surely only be a matter of time before some enthusiasts start to make EME contacts on the band, especially given that antenna size for a given gain is more manageable than on 6m. Perhaps all that is holding it back is the lack of sufficiently distant countries to work.)

The various other propagation modes to be found on 6m are in evidence on 4m, Sporadic E being responsible for much of the activity to date. In the

The current 4m distance record is 7543km, achieved during a QSO between EA6SX and ZS6WAB.

past the Sporadic E season offered little of interest to UK amateurs active on 4m. If anything, the opposite was the case, Sporadic E bringing with it strong carriers from TV and other commercial stations from around Europe. Nowadays much of that has gone and, instead, there is the opportunity to work DX, mostly out to one hop (meaning that signals are good, and simple equipment will more than do the job). Stations in Southern Europe (for example) can take advantage of trans-equatorial propagation to work over considerable distances on the band (the current distance record on the 4m website is EA6SX to ZS6WAB, a path of 7543km).

Meteor scatter is obviously available throughout the year when other propagation modes are not in evidence.

Over time we might expect more countries outside Europe to gain 4m allocations, at which time the DX chase will no doubt become more serious and very much more akin to what happens already on 6m.

In recent years there have been a number of portable operations activating rare squares on the band as well as activations by rare DXCC entities. For example, the Sovereign Military Order of Malta was active on 4m as 1A0KM in July 2007 during the Sporadic E season and again as 1A4A in October that year.

It is also worth noting that although US stations are not licensed for 4m, a beacon, WG2XPN, operates on 70.005MHz from locator square FM07fm. Trans-atlantic propagation has occasionally been observed on 4m (see sidebar on page 24), so it might be hoped that US and Canadian amateurs gain access to the band before too long.

SOURCES OF INFORMATION

This chapter has given a brief overview of the bands and covered some history. The rest of the book covers a wide range of topics from setting up your station for 4m and 6m, to operating techniques, contests and awards chasing and, perhaps most important of all, 4m and 6m propagation.

If you want to read beyond what is contained in these pages, there are many suitable sources. In terms of traditional publications, a number of excellent reference works are listed in the bibliography. Throughout the book there are also references to Internet sources. References pertaining to each chapter appear at the end of the chapter but Appendix B lists an extensive selection of sources.

Internet URLs are valid at the time of going to press, though it must be remembered that Internet sites are appearing and disappearing all the time, while some change their locations.

Surfing the net will take you to a huge number of useful sources; it would be quite impossible to list them all here. What this book will hopefully do is give you enough information to get you started and whet your appetite about aspects of Four and Six – whether it's propagation or antennas, history or DXpeditioning – that take your fancy and which you will want to explore further.

If you become a serious addict of Six you will almost certainly also want to join one of the 6m clubs such as the UK Six Metre Group. UKSMG publishes an excellent magazine, *Six News*, runs a very good website with lots of material available exclusively to members and organises social events where you can meet other enthusiasts and hear talks on DXpeditions and the like. The group also sells a CD containing all back issues of *Six News*. This is well worth a read, perhaps during those times when Six is quiet!

There is no equivalent 4m club in the UK, though The Four Metres Website **[4]** provides much useful information.

Elsewhere in the world 6m enthusiasts join together, for example at the regular G3WOS gatherings, to swap war stories and hear advice from some of the experts. It is worth tapping into these networks if you become serious about Six. Not only will they give you access to expert advice but

***Six News*, the excellent bi-monthly journal of the UK Six Metre Group.**

The crowd at a G3WOS 6m barbecue.

they are also a lot of fun. There are far fewer equivalent bodies or resources relating to 4m for the time being, but expect that to change as time goes on.

REFERENCES

[1] UK Six Metre Group: uksmg.org
[2] Six Metre Transceivers, *Six News*, Issue 65, May 2000.
[3] *Historique des Bandes de Frequences HF & VHF Disponibles aux Radioamateurs*: https://tinyurl.com/y6rdaums
[4] Four metres website: www.70mhz.org

...AND REMEMBER THE
VHF DXER'S OPERATING ETHIC:

1. If all the DX is S9, you're missing the real DX.
2. If the band is 'closed', find a way to open it!

(source: GM3SEK)

Chapter 2

Equipment & Station Design

MOST CURRENT TRANSCEIVERS include 6m as standard and an increasing number, particularly the European models, also include 4m. This makes it easier than ever to become active on one or both of these bands. 6m transverters are still available, either ready-made or in kit form, but it's unlikely that serious 6m DXers, for example, would choose that route nowadays. It is more likely that you may want to use a transverter for 4m operation although, again, most regular users have now set themselves up with a transceiver that covers the band.

6M EQUIPMENT

The choice of equipment very much depends on what sort of 6m operating you have in mind. If all you want to do is some 'ragchewing' with local amateurs, or you are looking to operate mobile, maybe working through a local repeater, then the main criteria may be minimum cost and small size. Neither of these should be a barrier nowadays. There are many excellent and small transceivers that cover a wide range of frequencies. Their new prices are modest considering the features you get for your money, and many perfectly adequate radios can be bought second-hand for a very reasonable outlay. You can buy a set specifically designed for channelised FM operation, such as the Yaesu FT-8900R, which is a quad-band (10, 6, 2m, 70cm) FM transceiver designed very much for mobile operation. Or you can have a multimode transceiver, most of which cover the HF bands as well as some of the VHF bands, good examples being the Yaesu FT-891, the Icom IC-7100 (which also covers 4m) and the Kenwood TS-480, which all cover HF and 6m with decent performance while being small and light enough for mobile or portable operation. If, on the other hand, you are looking for a high-performance solution to 6m operation for DXing and / or contesting, then you really need to take a different approach. Using a transverter is still a viable way to get on Six if you already have a good quality HF transceiver.

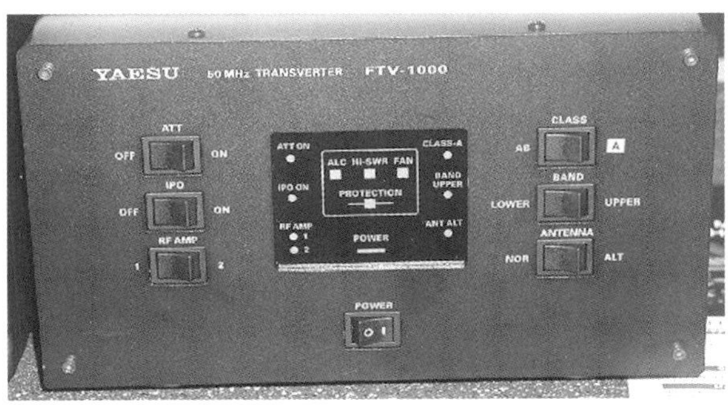

The Yaesu 6m transverter, designed to work with the FT-1000 series of transceivers and capable of 200W output.

Ten-Tec of the USA used to sell a 6m transverter kit to work with a 20m transceiver, for example, while Yaesu sold the FTV-1000, designed to work with the FT-1000MP MkV transceiver but capable of operation with other transceivers, albeit requiring its own power supply when not mated with the FT-1000MP. Some transceivers are more suitable than others for use with an external transverter – if you are considering this route to getting on Six then ensure that your transceiver has a low-level transverter output.

In the past many multimode VHF / UHF base station transceivers such as the Yaesu FT-736 also offered excellent 6m performance. Such radios no longer appear to be manufactured, the line between HF and VHF equipment now tending to be drawn above 50MHz. So, for example, the Icom-910H offered 2m and 70cm base station capability (with 23cm as an option) although, interestingly, its replacement, the IC-9100, covered everything from 160m through to 23cm (except 4m). However, Icom have reverted to offering the IC-7300 (and its 'big brother', the IC-7610) to cover the bands up to 50MHz (or 70MHz in the case of the IC-7300) and complementing that with a VHF / UHF transceiver, the IC-9700.

On both 6m and 4m, as on other VHF bands, a panoramic display is beneficial as it can be a handy way of alerting you to band openings. The Icom IC-756 series of transceivers has always been popular with 6m operators, partly for this reason and partly because they were one of the first mainstream transceivers to include 6m along with the HF bands. The IC-7610, which supersedes the '756 and '7600 series, has the same capability, being able also to drive an external monitor on which the main display is reproduced.

But perhaps the biggest change in this respect in the past few years is that SDR (Software Defined Radio) equipment such as that from FlexRadio Systems and Anan now tends to have coverage extending to 6m and above, although even mainstream transceivers tend to come with an inbuilt display, usually one that can also be fed to an external monitor for even better resolution. The FTdx101, TS-890, IC-7610 and IC-7851 all have this capability, for example, as will the new Elecraft K4. And many other transceivers have an output in front of the main roofing filter, allowing connection to an

The Apache Labs Anan-7000DLE SDR transceiver for HF and 6m.

external panoramic display such as the LP-Pan from TelePost Inc or the Elecraft P3.

What other considerations apply? If you are intent on DXing, then you will really need to be able to operate CW (brush up your CW if it's not a mode you currently use) and / or WSJT. So, you may want CW filters in your transceiver where these are an after-purchase option. Increasingly, though, add-on CW filters are a thing of the past – nowadays most transceivers use digital signal processing, at least for part of the receive chain, and offer a range of bandwidths suitable for CW reception. A general-coverage receive capability can be useful so that you can listen outside the band for commercial signals, which may give an indication of potential band openings. Most transceivers nowadays have this capability but SDR transceivers can be especially useful as you can literally 'watch' several pieces of spectrum below and up to 50MHz and, by doing so, see signals start to appear as the MUF increases (for example, during the onset of a Sporadic E opening). A dual-watch receive capability is also useful for monitoring both sides of a DX pile-up.

On the transmit side, the question of power is an obvious one. Most of the base station and even mobile radios will now generate 50 or 100 watts on 50MHz (200W from the TS-990, FTdx101MP and IC-7851), which is more than adequate to get started on the band and to exploit many of the Sporadic E openings, during which signal strengths can be very high. Those transceivers that include 70MHz will normally put out 50 to 100 watts on the band. But for long-haul and / or weak-signal operation it makes sense to be able to run as much power as your licence allows. This varies from country to country. In the US, for example, the power limit on 6m is 1.5kW whereas in the UK it is 400 watts at the antenna. This means that you are allowed to take feeder losses into account when considering what power you can run out of the transmitter. This said, it doesn't make sense, as will be discussed later, to accept high feeder loss and make up for it with additional transmitted power; this won't help on receive where the feeder loss will attenuate those weak incoming signals. But this advice from Ian White, GM3SEK, is equally pertinent, "A bit more ERP will boost the station's capabilities into a higher league, where long-distance 'scatter' QSOs can be made almost any time, especially when the band is nearly open but not quite. If signals fade, just a one-way trace of signal from the station with higher ERP is often enough to keep the QSO alive until it can be completed. In difficult conditions, this can increase the completion rate quite dramatically. Given that most 6m users in the UK would not be able to increase their antenna size significantly, and the receive side is limited by external noise, then the power amplifier is the only way forward. Given also that most rigs can already deliver 100W, amplifiers of less than the 400 – 500W class are largely a waste of time and money."

At the time of writing, the main options at the high end are the TS-890 from Kenwood, the FTdx101 from Yaesu, the K4 from Elecraft and the

A selection of high-end transceivers offering 6m capability: the Kenwood TS-890 (top); the Yaesu FTdx101 (centre); and the Icom IC-7610 (bottom).

IC-7610 and IC-7851 from Icom. But the SDR radio manufacturers have some exceptional offerings too – Anan, FlexRadio and the SunSDR2 are good examples. And if you are thinking of operating your station remotely at some time in the future, the SDR rigs make this particularly easy although most main station transceivers are also geared up for remote operation nowadays.

In summary, there is no single piece of advice that applies to the choice of a radio, as your selection will very much be determined by what you want. Do you operate portable and / or mobile? Do you want to be able to operate on other bands or only on 6m (and perhaps 4m)? Is it important to you to be able to monitor commercial frequencies outside the amateur bands in order to spot likely band openings? What modes are likely to be of most interest to you? And so on, including perhaps the most limiting criterion of all – what is your budget? The best way is to make a checklist of criteria and match it against the various radios on the market. And, where possible, track down the reviews that have been published by reputable bodies such as the ARRL and the RSGB. These will include laboratory measurements of performance and you will need to check out the 50MHz parameters. But it's probably fair to say that, nowadays, the performance of most transceivers, even at the budget end, is pretty good, so the ergonomics may well be of just as much importance in your selection.

As a final comment on transceivers, it is worth noting the growing trend, already alluded to, to use an SDR (Software Defined Radio) receiver in conjunction with your main transceiver (or to use an SDR transceiver). There are several benefits. One is to be able to access the whole, or a significant part, of the band, which can then be presented in graphical format on your PC screen via software such as WinSDR **[1]**. This allows a bandscope display of the frequency segment being monitored, so that you can see immediately if signals are being received. Use it also with software such as CW Skimmer **[2]** and you will even see the callsigns of stations being copied. If the SDR receiver is fed after the roofing filter in your transceiver much of the benefit is lost, as you will then only be able to display and monitor 3kHz of the band. As a 6m enthusiast you really want a capability not only to monitor quite a lot of spectrum, but for this to go well below 50MHz, so that you can be alerted to potential band openings. Having said all of that, one of the 'most wished for' capabilities nowadays is probably to be able to monitor both 50.313MHz and 50.323MHz simultaneously, which could, for example, lead you to buying a radio with two receivers.

An interesting postscript to the topic of SDR is the availability of cheap SDR devices in USB form. These range from TV tuners sold for just a few pounds on eBay and elsewhere (The Realtek RTL2832U Radio & TV Dongle, for example) to products like the FUNCube Dongle **[3]**, designed by Howard Long, G6LVB. The former requires some additional software to be loaded to be usable as a broadband SDR receiver (just do an Internet search for details); the latter requires no additional drivers at all. Neither device is comparable to a full-spec SDR receiver or transceiver, but both offer the opportunity to monitor a wide range of frequencies easily and conveniently for relatively low cost.

The FUNCube Pro Dongle is an SDR in USB form, requiring no drivers.

4M EQUIPMENT

For many years the route UK amateurs took to become active on 4m was to convert PMR equipment (for AM or FM) or, in the case of SSB and CW, to build, usually a transverter rather than a 4m transmitter as such (although it was not uncommon to have a receive converter linked to, usually, a 10m receiver while, at the same time, using a separate transmitter specifically designed for Four). As was mentioned earlier in this book, there were one or two pieces of commercial equipment designed specifically for the band, but the selection was very limited.

Until recently, no multiband transceiver has included a true 4m capability, although the Yaesu FT-847, as sold in the UK, had a limited capability on Four. That is to say, it can run about 20 watts out but with poor efficiency and is generally considered somewhat 'deaf' on receive. A number of modifications exist on the Internet to improve both receive sensitivity and transmit efficiency and power output and at least one of the major UK distributors of Yaesu equipment offers to install their own version of the transmit modification for a modest charge, thereby achieving higher transmit power output for less internal heat dissipation, as well as improving the receive side.

Nowadays a few transverters are still manufactured, invariably solid-state and generally running at about the 25 watt level. One example is the Spectrum Communications product, which is available in versions with a 2m or 10m IF. With the release of the band to other European countries, similar products are now being manufactured elsewhere. If you are prepared to build your own, a number of designs appear, for example in the *RSGB VHF / UHF Handbook* **[4]** and OZ2M **[5]** has published a design which is available in kit form.

The Icom IC-7300, an excellent competitively-priced transceiver with both 6m and 4m capability, at least in the version sold in Europe.

At the high end, Kuhne now offer a high-performance transverter for the 4m band, priced (at the time of writing) at 369 euros.

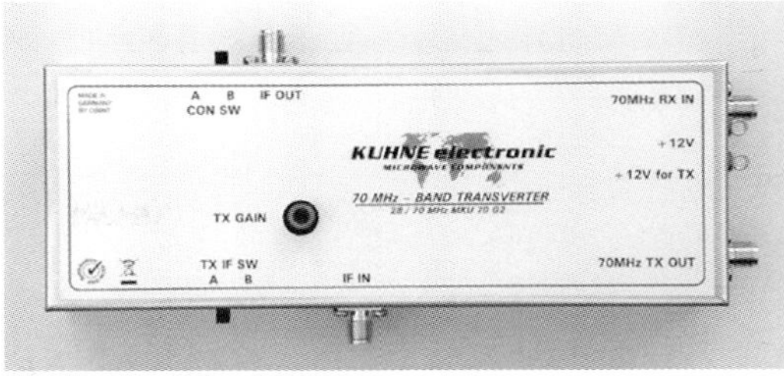

4m transverter from Kuhne of Germany.

But the big change in recent years has been the inclusion of 4m as standard on a number of mainstream radios. Icom were first out of the blocks with the IC-7100 but their best-known implementation is on the IC-7300, if only because that transceiver has sold in record numbers. There is no 70MHz button as such on the '7300, but the European version includes the band as standard. Kenwood have taken a similar approach with the TS-890S and Yaesu with the FTdx101.

The Wouxun KG-UV8H handheld covering 4m and 2m FM in the UK.

The latest FlexRadio 6000 series of equipment also has coverage extending above 70MHz and offers some extremely interesting possibilities, with the ability to monitor multiple blocks of spectrum simultaneously. The downside is that the price reflects the very high specification.

As a final comment on gear, it is a shame that Icom decided to launch the IC-705 without 4m capability but, in contrast, shortly before going to press, the Wouxun KG-UV8H dual-band handie-talkie was launched in the UK with 4m and 2m FM capability. It seems not unreasonable to expect that in the coming years there will be other new radios with 4m capability, at least in those versions aimed at the European market.

LINEAR AMPLIFIERS

As with transceivers, nowadays there are many linear amplifiers that operate on 6m and some for 4m or for both bands. Some are single-band units, designed specifically for Six. US companies such as Henry have manufactured these in the past and Alpha is currently offering the 8406 model. M2 manufactured a solid-state unit, which could prove to be ideal for portable operation, although this appears not to be made any more. The Lunar-Link Systems LA-62A by K1FO (SK) and the Commander 1200 from Command Technologies are two other popular choices.

Others are multiband, usually HF amplifiers with a 6m capability. There are many examples, both valve amplifiers such as the Acom 1000, and semiconductor units such as the Acom 600 and 700 (see below), Yaesu VL-1000 Quadra and the SPE Expert range of high power amplifiers. This list is by no means exhaustive – as with transceivers, we are increasingly seeing amplifiers that cover both HF and 6m.

New models covering 6m are coming on the market all the time as the popularity of 6m continues to grow and given that many of the output devices used are perfectly capable of working at 50MHz or above. OM Power, for example, has the OM3006+ valve amplifier, its predecessor reviewed by

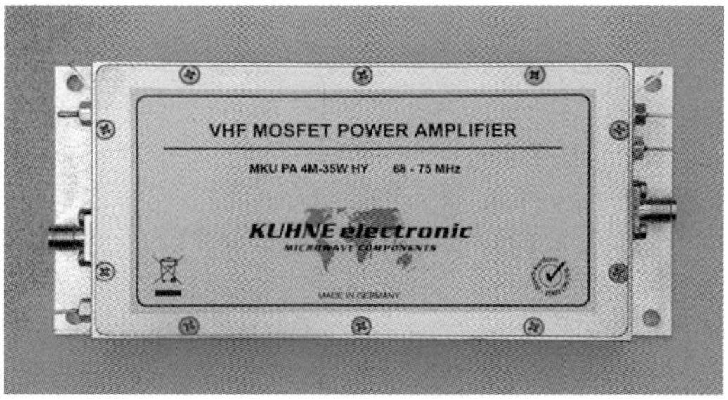

The Kuhne MOSFET amplifier for 70MHz.

The Gemini 4 linear amplifier covering both 6m and 4m.

G4BWP / A65BD and G8VR in *Six News*, Issues 110 and 111. The OM2000+ covers 160 through 6m.

Linear Amp UK manufacture the Gemini 4 amplifier, which covers both 6m and 4m with an output of 300W. This may be the only commercial amplifier offering 70MHz capability, perhaps because high power has not really been a requirement when most potential 'DX' is within a single hop. But with DX now becoming available from the Middle East, for example, this is no doubt set to change.

Acom has several amplifiers covering both HF and 6m, including the Acom 1500, an earlier version of which was released in 2012 and reviewed by Peter Hart, G3SJX, in the December 2012 *RadCom*. Their new 700S amplifier is a solid-state unit with 700W output (its predecessor, the Acom 600S was reviewed in *QST* in August 2015).

Incidentally, of the Expert range, the Expert 2K-FA, while a great amplifier, has to be derated when running data modes, whereas its smaller brother, the Expert 1.5K-FA does not. So, it pays to fully understand what you are buying when looking at amplifiers.

Acom's new 700S solid-state amplifier covers 6m.

The OM Power OM2000+ valve linear amplifier.

The Expert 1.5K-FA solid state HF + 6m amplifier.

Any of these amplifiers will be perfectly suitable for 6m operation, the main trick with amplifiers being to avoid overdriving them, so it always pays to monitor your signal or have a local amateur check it for you (but if you are very loud with him, it is important that he can differentiate between a poor signal on your side and simple overloading of the receiver on his side). Generally speaking, it is worth buying an amplifier capable of producing more power than you will actually want to run, so that you can always operate it well within its specification, for cleanest signal and longest life. And if you intend to operate data modes such as WSJT, the high duty cycle again means that the amplifier should be over-rated to avoid early failure. Building your own amplifier is a major undertaking, but designs do appear in various places, including the *RSGB VHF / UHF Handbook* **[6]** and there is a very popular design for an expedition amplifier on the G3WOS web page **[7]**.

Specifically for 4m, there are some ex-PMR commercial models (the solid-state A200 being one such) which are easily converted to 4m and provide useful power levels of tens of watts.

ANCILLARY EQUIPMENT

In terms of other equipment, this will depend on circumstances. A second (general coverage) receiver to allow you to check for spurious emissions and generally to monitor your transmitted signal is always a good idea.

Your transceiver may well have a built-in SWR meter, which will allow a basic check on whether the antenna is working correctly. Otherwise an external SWR / power meter is a valuable asset in the shack. If you are going to do anything more than put up a single Yagi you may also want some more specialist antenna test equipment, for example a swept analyser such as the NanoVNA, miniVNA or the RigExpert AA-55 (which works up to 55MHz) or AA-230 (up to 230MHz). Such an analyser will not only allow you to check antenna SWR and impedance, but also feedline loss, and will allow you to cut phasing lines, for example, when building a stacked array.

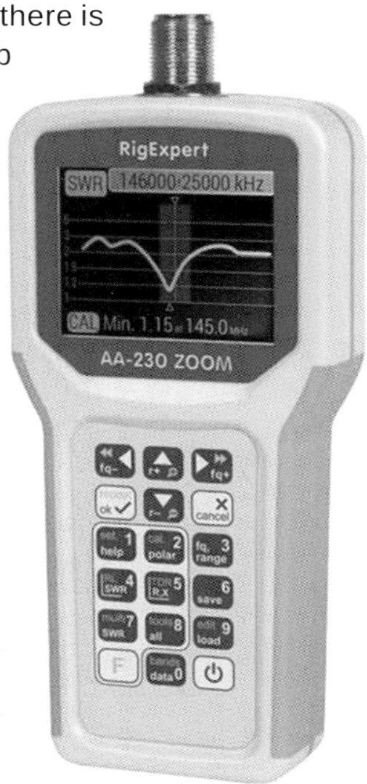

The RigExpert AA-230 antenna analyser, usable up to 230MHz.

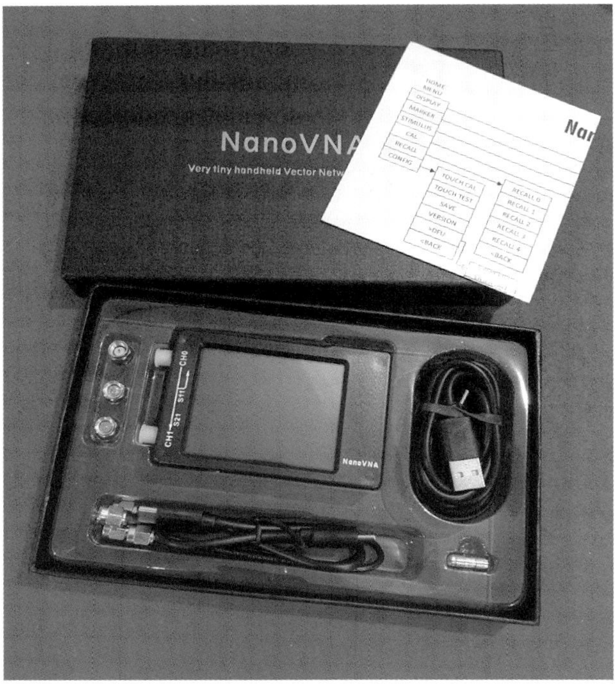

The low-cost NanoVNA, able to make a wide range of measurements up to 900MHz.

Other equipment that you might want to consider includes filters of various descriptions. In the shack you may wish to install a low pass filter on the output of your transmitter (or linear amplifier if the filter can handle the power) to reduce any harmonic radiation to a minimum (but do ensure it is designed for 50MHz or 70MHz operation). Designs appear in a number of places including the *ARRL Handbook* and on the Internet **[8]**.

The other use for filters is when you have EMC problems with a neighbour's electronic device, at which time you may need to install one or more varieties of band-stop filter, choke etc on the various ports of his equipment (antenna lead, power line, etc). This topic is further discussed later in this chapter under EMC.

COMPUTERS AND SOFTWARE

No shack is complete nowadays without a PC. The PC has many uses and this is not the place to describe them in detail. They are covered in some depth in the RSGB book *Computers in Amateur Radio* **[9]** for example, but also in many other places.

Briefly, there are two areas of use, local and the wider network. Locally, you will use the PC for logging, for antenna modelling, for running software such as propagation forecasts (though this is not too useful at 50 or 70MHz) and, increasingly, as a bandscope in conjunction with an SDR receiver (or, indeed, as a key element in a station designed around an SDR transceiver). 'Logging' is something of a misnomer, as most so-called logging programs nowadays are really 'station management' programs, handling a wide variety of tasks, including such things as rotator control, Cluster interface, Logbook of The World submission, award tracking etc.

4m enthusiasts should be aware that many logging programs do not recognise the 4m band, so should bear this in mind when selecting logging software. The same applies to contest logging software. For RSGB contests, SDV (the VHF version of the EI5DI's SD software) does the job nicely, but is by no means the only option available, with MINOS by Mike Goodey, G0GJV, being very popular for RSGB VHF contests. You will almost certainly also want to use the PC for running data modes software, especially WSJT, described in the specialist modes operating section of this book. And you will want to be able to access the PSKReporter, RBN and Cluster websites as well as chat rooms such as that run by ON4KST.

Most SDR transceivers do the majority of their own processing with a simple control program running on the PC but the lower-end SDR radios require the PC to do the heavy lifting, so this may be a factor to bear in mind. What does all this mean for the PC's specification? For use for Internet access and basic logging functions a PC retired from other functions within the home may be perfectly adequate. For use as part of an SDR capability, a fast PC with high-specification soundcard will be required. The good news is that the price of such computing capability has fallen dramatically in recent years. As far as the operating system is concerned, most software for amateur radio use is designed to work with the Windows operating system as this is by the far the most ubiquitous. But beware that, because much of this software has been designed by fellow amateurs, they may not always have time to upgrade it to work with the very latest version of Windows, so it is always worth checking.

Some pieces of ancillary equipment, including transceivers themselves, still use a serial port rather than a USB connector, whereas very few modern PCs have a serial interface, so an adapter will be required. Some work better than others so beware which chipset and driver program your USB / serial adapter uses – those using the FTDI chipset, although more expensive than some others, appear to be the most reliable. There is a lot of useful information available on the Internet if you run into difficulties.

It is worth mentioning that WSJT will run well on Raspberry Pi, so you can have a cheap system dedicated to WSJT operation. That said, there is a case for being able to run multiple instances of WSJT and / or JTDX, perhaps to monitor more than one band or channel, in which case a fast PC may be of value. WSJT will also run on Linux or on OS-X (Macintosh).

The WSJT and JTDX programs are covered in detail in Chapter 6. But recent years have seen the arrival of lots of other software to enhance our operating, record keeping, station management and more. While these programs are not specific to 6m or 4m, it is worthwhile giving them at least a mention here. Further information on any of them can, as always, be obtained from the Internet, often including 'how to' videos on *YouTube*.

Logging

First, logging programs. There are many to choose from, both for station management and for contest logging. For station management, Logger 32 is a good example of a logging program that will interface to your PC, to the Internet (for RBN, Cluster, upload to Logbook of the World and so on) and more. Logger 32 can link up with WSJT-X and JTDX via UDP in order to log QSOs as they are made and to generate a kind of band map showing the digital stations recently decoded by WSJT-X / JTDX. The process is explained in detail in Gary's, ZL2IFB, documentation **[10]**.

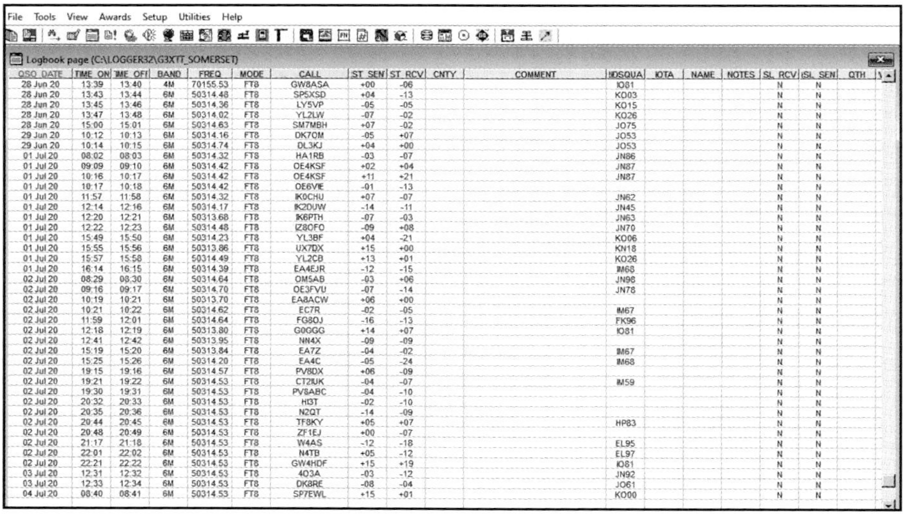

QSO DATE	TIME ON	ME OFF	BAND	FREQ	MODE	CALL	ST SEN	ST RCV	CNTY	COMMENT	IDSQUA	IOTA	NAME	NOTES	SL RCV	SL SEN	QTH
28 Jun 20	13:39	13:40	4M	70155.53	FT8	GW8ASA	+00	-06			IO81				N	N	
28 Jun 20	13:43	13:44	6M	50314.48	FT8	SP5XSD	+04	-13			KO03				N	N	
28 Jun 20	13:45	13:46	6M	50314.36	FT8	LY5VP	-05	-05			KO15				N	N	
28 Jun 20	13:47	13:48	6M	50314.02	FT8	YL2LW	-07	-02			KO26				N	N	
28 Jun 20	15:00	15:01	6M	50314.63	FT8	SM7MBH	+07	-02			JO75				N	N	
29 Jun 20	10:12	10:13	6M	50314.16	FT8	DK7OM	-05	+07			JO53				N	N	
29 Jun 20	10:14	10:15	6M	50314.74	FT8	DL3KJ	+04	+00			JO53				N	N	
01 Jul 20	08:02	08:03	6M	50314.32	FT8	HA1RB	-03	-07			JN86				N	N	
01 Jul 20	09:09	09:10	6M	50314.42	FT8	OE4KSF	+02	+04			JN87				N	N	
01 Jul 20	10:16	10:17	6M	50314.42	FT8	OE4KSF	+11	+21			JN87				N	N	
01 Jul 20	10:17	10:18	6M	50314.42	FT8	OE6VIE	-01	-13							N	N	
01 Jul 20	11:57	11:58	6M	50314.32	FT8	IK0CHU	+07	-07			JN62				N	N	
01 Jul 20	12:14	12:16	6M	50314.17	FT8	IK2DUW	-14	-11			JN45				N	N	
01 Jul 20	12:20	12:21	6M	50313.68	FT8	IK6PTH	-07	-03			JN63				N	N	
01 Jul 20	12:22	12:23	6M	50314.48	FT8	IZ8OFO	-09	+08			JN70				N	N	
01 Jul 20	15:49	15:50	6M	50314.23	FT8	YL3BF	+04	-21			KO06				N	N	
01 Jul 20	15:55	15:56	6M	50313.86	FT8	UX7DX	+15	+00			KN18				N	N	
01 Jul 20	15:57	15:58	6M	50314.49	FT8	YL2CB	+13	+01			KO26				N	N	
01 Jul 20	16:14	16:15	6M	50314.39	FT8	EA4EJR	-12	-15			IM68				N	N	
02 Jul 20	08:29	08:30	6M	50314.64	FT8	OM5AB	-03	+06			JN98				N	N	
02 Jul 20	09:16	09:17	6M	50314.70	FT8	OE3FVU	-07	-14			JN78				N	N	
02 Jul 20	10:19	10:21	6M	50313.70	FT8	EA8ACW	+06	+00							N	N	
02 Jul 20	10:21	10:22	6M	50314.62	FT8	EC7R	-02	-05			IM67				N	N	
02 Jul 20	11:59	12:01	6M	50314.64	FT8	FG8OJ	-16	-13			FK96				N	N	
02 Jul 20	12:18	12:19	6M	50313.80	FT8	G0GGG	+14	+07			IO81				N	N	
02 Jul 20	12:41	12:42	6M	50313.95	FT8	NN4X	-09	-09							N	N	
02 Jul 20	15:19	15:20	6M	50313.84	FT8	EA7Z	-04	-02			IM67				N	N	
02 Jul 20	15:25	15:26	6M	50314.20	FT8	EA4C	-05	-24			IM68				N	N	
02 Jul 20	19:15	19:16	6M	50314.57	FT8	PV8DX	+06	-09							N	N	
02 Jul 20	19:21	19:22	6M	50314.53	FT8	CT2IUK	-04	-07			IM59				N	N	
02 Jul 20	19:30	19:31	6M	50314.53	FT8	PV8ABC	-04	-10							N	N	
02 Jul 20	20:32	20:33	6M	50314.53	FT8	HI3T	-02	-10							N	N	
02 Jul 20	20:35	20:36	6M	50314.53	FT8	N2QT	-14	-09							N	N	
02 Jul 20	20:44	20:45	6M	50314.53	FT8	TF8KY	+05	+07			HP83				N	N	
02 Jul 20	20:48	20:49	6M	50314.53	FT8	ZF1EJ	+00	-07							N	N	
02 Jul 20	21:17	21:18	6M	50314.53	FT8	W4AS	-12	-18			EL95				N	N	
02 Jul 20	22:01	22:02	6M	50314.53	FT8	N4TB	+05	-12			EL97				N	N	
02 Jul 20	22:21	22:22	6M	50314.53	FT8	GW4HDF	+15	+19			IO81				N	N	
03 Jul 20	12:31	12:32	6M	50314.53	FT8	4O3A	-03	-12			JN92				N	N	
03 Jul 20	12:33	12:34	6M	50314.53	FT8	DK8RE	-08	-04			JO61				N	N	
04 Jul 20	08:40	08:41	6M	50314.53	FT8	SP7EWL	+15	+01			KO00				N	N	

Logger 32 integrates with WSJT to allow seamless logging of FT8 QSOs.

Winlog32, Log4OM, Ham Radio Deluxe and others work well under Windows but, of course, there are also logging programs available for Mac or Linux.

As far as contest logging is concerned, N1MM+ is probably the most common, followed by WriteLog for Windows and others such as Win-Test. N1MM+ integrates nicely with WSJT with lots of online help available, as also does WriteLog.

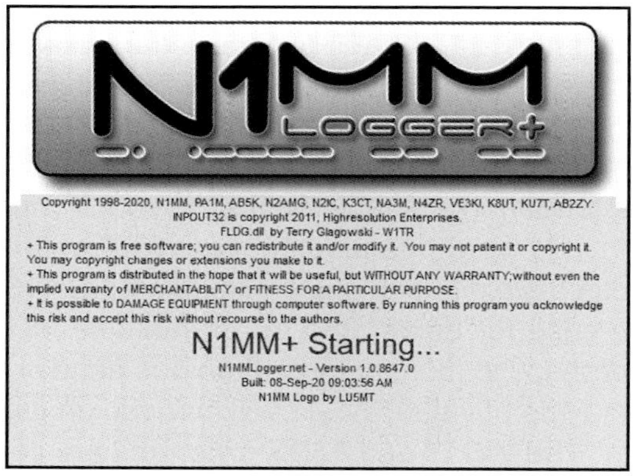

N1MM+ is probably the most commonly-used contest logging program and it also integrates well with WSJT.

Grid Tracker

Grid tracker [11] is a handy utility working with WSJT or JTDX, collecting QSO data and displaying in graphical (map) format.

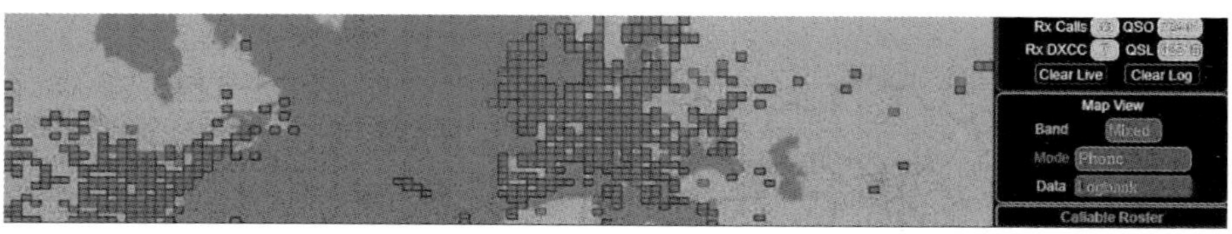

A screenshot from Grid Tracker.

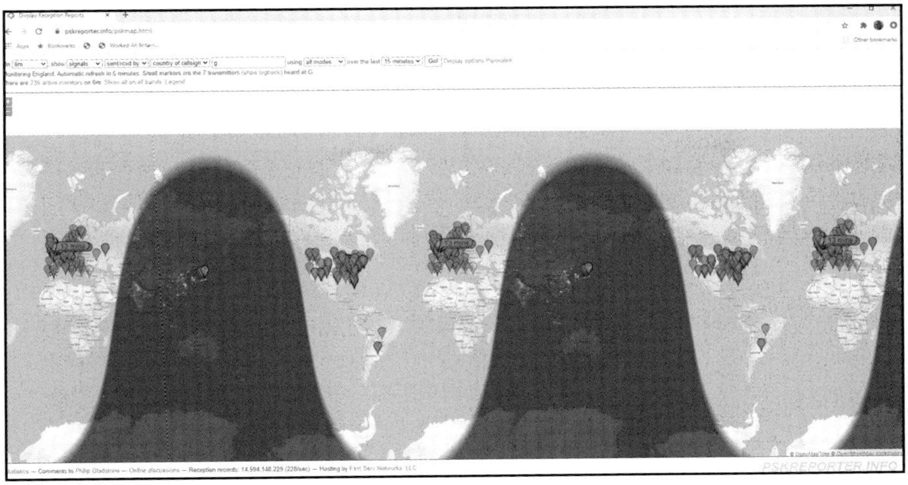

A screenshot from the PSKReporter website.

PSKReporter

It is worth at this point mentioning the PSKReporter website. This displays signal reports of digital signals by band and mode and is an excellent tool for keeping an eye on band conditions and knowing how your own signals are being received elsewhere. It covers all the WSJT modes, for example, so you can see where QSOs are being made. But it can also be useful to search on your own callsign and see where you are being heard.

JT Alert

It's also worth mentioning another handy piece of software. While WSJT can be integrated directly with the Logger 32 logging program, JT Alert **[12]** by VK3AMA allows it to interface to a number of other logging programs. But it does much more than that, tracking your progress towards awards and letting you know, for example, when a spotted station is a new one for you. For integrating WSJT and JT Alert there is lots of online help, including *YouTube* videos.

THE INTERNET AND TELEPHONE ALERTING

It may seem obvious, but nowadays you will want the Internet to be integrated into your station. There is so much that it puts at your fingertips, from the DX Cluster system to the ON4KST chat room, from real-time propagation data to world locator square maps, and much more. The use of these tools is described in detail in Chapter 5.

But simply having the Internet on a screen by your transceiver may not be enough. After all, do you spend all your time in the shack? It has been said many times in this book and elsewhere that 6m and 4m openings can be fleeting and, Murphy's Law being what it is, will invariably occur just as you have gone downstairs, into the garden or to the shops. So, it is worth

G7VJR's DXLite, here showing 4-metre 'spots', is designed to be monitored on a Smartphone.

thinking about how to deal with these situations. In the past, 6m DXers would have shared lists of what they needed and, if a 'Wanted' one had appeared, would use the phone to alert each other to the DX. This may well still happen on a localised basis but there is the general assumption that everyone has the Internet nowadays and that other forms of alerting are unnecessary.

What to do? A good start is to ensure that, when there is a rare one scheduled to be on, you have the Internet with you wherever you go. If you have a wi-fi network in the house this should be easy. Use a laptop and carry it from room to room. Your wi-fi network may even extend to the garden. Or for ease of use, carry a wi-fi capable device in your pocket (almost certainly a Smartphone of some sort) and connect via one of the fast Cluster-like services such as DXLite server developed and run by Michael Wells, G7VJR, designed with PDA and mobile phone users in mind **[13]**.

The other way to be alerted while you are out and about is to have the Internet, working in conjunction with your logging software, alert you by way of a text (SMS) message to your mobile phone or a transmission to your VHF handheld. Many logging software packages have the capability of doing either or both of these things, to a set of predetermined criteria (e.g. a new one on the band, or a new mode slot on the band).

Finally, you may want to arrange with friends for mutual alerting. WhatsApp, Messenger, Skype and other messaging applications are your friends in this case.

DEALING WITH EMC PROBLEMS

Although Band 1 TV is no longer an issue in most parts of the world, anyone planning to chase weak-signal DX on 6m or 4m will almost certainly be faced with EMC issues, particularly when running high power. The most common problem is with masthead and other distribution amplifiers for FM radio and TV. Many of these are broadband in their characteristics, exhibiting significant gain at 50MHz. The solution has to lie at the neighbour's location when such problems occur.

After output filtering, there is nothing that you can do in your own shack, other than reducing power, which is unacceptable if you are to be ready to chase DX when those short openings occur. The solution may require several elements and each case will be different. The sort of options to investigate include:

- Dealing with common-mode signals (induced directly into the feeder) at the input to the amplifier. This can be tackled by installing common-mode chokes in series with the input to the preamplifier. Such a choke can consist of 12 turns of miniature coax wound on Fair-Rite 2643802702 cores, as described, for example, in the *RSGB Yearbook*.
- Dealing with differential-mode 50MHz signals (picked up by the TV antenna) at the input to the amplifier. One source recommends a lowish-Q parallel resonant circuit (500nH and 20pF) in-line between the common-mode choke and the input to the amplifier. This has over 30dB attenuation at 50MHz but less than 3dB attenuation at 90MHz.
- Dealing with 50 or 70MHz signals reaching the amplifier via the DC voltage feed. The solution to this is a common-mode choke on the DC voltage line.

A common-mode choke made from turns of coax wound on to a ferrite core.

You may find your transmitted signals getting into other equipment, and similar advice applies but, in the UK, masthead and distribution amplifiers are by far the most common problem. Unfortunately, they require the neighbour to be co-operative, given that the solution has to be at his location. In those countries where cable TV is widespread, the major EMC problems with neighbours are likely to arise from other consumer devices such as stereo systems, computer speakers and the like. These are usually indicative of inferior design of the product and may not be simple to eliminate. In general, though, EMC problems to neighbours are much less of a problem than they were in years past.

You may also suffer incoming EMC problems if you are in a noisy location, preventing you from pulling through the weakest signals. Dealing with these requires a portable receiver and the patience to walk the neighbourhood to track down the offending device(s). The solution will depend entirely on what it is you find and, again, co-operation from the neighbour(s) concerned.

Remember, though, that a major contribution to reducing noise can be achieved simply through having a good antenna with narrow beamwidth, which will go a long way to reducing incoming noise (except, of course, when you have to beam in the direction of the noise source!) Stacking Yagis, as well as improving gain and take-off angle, is also a great way of improving your signal-to-noise ratio.

Costas, SV1DH, who lives in a noisy urban environment, has experimented with an ANC-4 wide-band noise canceller (using a 5-element Yagi as the noise antenna) but not found this effective most of the time. Such a system can suppress the noise from a single source by up to 20dB, but often in such an environment one is faced with multiple noise sources. Some 6m DXers report good results with the DX Engineering NCC-1 Variable

Phasing Controller. Other custom-made noise reduction systems are available, such as the MFJ-1025 and MFJ-1026. They work up to a point, provided you can get a big enough signal on the noise antenna to null out the noise being received on the main antenna. However, they are not so good when dealing with multiple noise sources. But it is also worth noting that modern SDR rigs (those with at least two ADCs) can provide excellent noise reduction. The Flex and Anan series of radios are a case in point and there is an excellent *YouTube* video **[14]** by G7CLF demonstrating noise reduction with an Anan 100D.

DEALING WITH NOISE SOURCES

HF operators are used to living with high levels of background noise, both natural and manmade. VHF operators expect to have low noise levels and to be able to deal with extremely weak signals. But, increasingly, man-made noise is becoming a problem in many locations. There are several ways of mitigating the effects, although very few operators will be able to escape the problem entirely.

One solution, to which an increasing number of operators are turning, is to have a remote station in a rural (quiet) location, which they can run from home. Remote operation is covered elsewhere in this book, as it has other uses too (for example, being able to run your home station when out and about).

But what to do if you are faced with man-made interference at your main station and want to be able to work DX from there? Sometimes the interference is from specific directions which are of little or no importance. Life being what it is, though, it will almost certainly emanate from one or more key directions!

One 6m DXer to have made a specific study of the matter is Chris, G3WOS. Chris found that the increasing levels of noise around his home were playing havoc with his 6m DX activities, particularly EME where signals can be very weak indeed and, when trying to exploit ground gain, it is necessary to work at the lowest take-off angles (and therefore be beaming right through the surrounding urban area). Although there has been a lot of talk about the problems of power line adapters (for datacomms distribution within houses over mains wiring), Chris found that the worst offenders were switched-mode power supplies, plasma TVs and low-voltage downlights. These various sources can often be tracked down using a portable receiver and simply walking the neighbourhood. The problem then becomes one of persuading the offending household(s) that there is a problem that needs to be attended to. This is where many amateurs find it hard to take the next step as it involves diplomacy rather than technology. It's really a matter of whether the problem is affecting your DXing enjoyment to the extent that drastic measures are called for.

In reality, if approached the right way, many people are prepared to listen and to cooperate, and the solutions are often straightforward and can be portrayed as helping to solve a potentially dangerous situation (for example, an enclosed power supply that may be overheating and therefore potentially a fire hazard).

Chris gives some valuable hints and tips (along with case studies) on his own website **[15]** and also refers to ON4WW's website **[16]** which, again, contains lots of useful information. And UK amateurs should also remember that the RSGB EMC Committee is there to help. There is often a Committee member living not too far away who can tap into a wealth of experience and advice. The moral of the story is to persevere in dealing with such issues, rather than allow them to adversely impact your enjoyment of the hobby.

REMOTE OPERATION

Given the difficulties in building a really competitive 6m or 4m station in many home locations (surrounding hills, lack of space, planning limitations, high local noise levels, EMC problems etc), a number of serious 6m and 4m DXers have looked for a workaround by setting up a station elsewhere. This may be at a friend's location or one owned by the amateur himself but remote from the main home.

One well-known UK 6m DXer, for example, has an old World War II bunker where he has installed his main 6m station, not too far from his home but on a much more advantageous site. The benefits of such an approach are obvious if it results in a more effective station but there are disadvantages too, usually

The MFJ-1234 RigPi box allows almost all transceivers to be interfaced to the Internet for remote operation.

in terms of the time it takes to get to the site when a DX station appears. One way around this is to arrange to be able to operate the remote station from elsewhere, either from your home or from your work. The benefits also work in the reverse direction too. There can be nothing more irritating than being stuck at work or at a friend's or relative's house and knowing that you are missing a rare one on Six! But many 6 and 4m DXers nowadays get round this by operating their home station remotely, to avoid missing those fleeting openings.

Remote operation is much easier nowadays than in years gone by. The ubiquity of the Internet offers a ready-made networking solution, always on, and many transceivers nowadays come ready to interface to the Internet. Good examples are the Kenwood TS-890, the Elecraft K3, the IC-7610, the FTdx101 (with external networking interface) and, perhaps most obviously, various Software Defined Radios such as the various Flex models, which are not simply built with a networking capability but have the computer interface as an inherent aspect of their design and operation.

The MFJ-1234 RigPi box allows pretty much any transceiver to be interfaced to the internet. And the various boxes from remoterig.com offer a wide range of solutions.

Remote station operation, though, requires more than being able to control the transceiver remotely. You need to be able to key the radio for CW or send and receive voice for SSB. You also need to be able to rotate the antenna and control the amplifier. Several articles have dealt with these aspects (e.g. **[17]**) and if you intend to go down this route you would be well advised to consult some of those sources and do some thorough planning based around your specific requirements.

But don't be put off as remote operation is getting easier all the time, and there are now several vendors who supply the equipment you will need at either end of the link in order to control not just your transceiver but other items such as rotator and linear amplifier. There is no doubt, though, that remote operation of data modes can be much simpler to set up than for CW or SSB. There is lots of information available on the Internet and several books (for example **[18]**) covering this increasingly important and accessible topic.

A final word of warning on remote operation, though. Do ensure that you set up the capability in a way which is within the terms of your licence

Another solution for remote operating is offered by remoterig.com

(licensing authorities tend to be nervous about remote operation because there is always the potential for a carrier to be transmitted under fault conditions with the remote operator unable to turn off the transmitter). Also, the rules and regulations pertaining to certain awards and contests rule out remote operation for those particular activities, so do please check that you fall within the relevant restrictions.

OTHER CONSIDERATIONS

There is much more that could be said about designing and building your station, but they are considerations that are by no means specific to 6m and 4m operation. Subjects such as station layout for operating convenience and comfort. Safety concerns, both electrical (for example, an easy-to-access main switch) and mechanical (particularly where antenna masts are concerned). Cable routing for minimal noise pickup. Good earthing. Station security and insurance. And so on. These are covered in other sources, for example in *The Amateur Radio Operating Manual*. In the present context, a significant factor may be whether your station is engineered purely for 6m and / or 4m, or whether you operate other bands, too. If the latter, is your 6m and / or 4m ready to go, given that band openings can be fleeting and happen at short notice, or is a certain amount of station reconfiguration required?

But perhaps the most pertinent piece of advice is to keep a station notebook and record details not only of station wiring and configuration (plus, perhaps, equipment serial numbers and the like) but also make a note of key operating parameters such as cable loss. This is especially important in weak-signal VHF work. If you are able to characterise your feeders and then re-measure them on, say, an annual basis, you will quickly see whether their performance is suffering and they need to be replaced. Moisture ingress in particular can be insidious and not visible to the eye, but can seriously affect cable loss.

Connectors can also deteriorate over time, especially if the connections were not well made or sealed in the first place. For VHF use, it is never a good idea to buy poor quality connectors – money spent on good ones (probably the pressure-seal type) is money well spent. And any connections should also be well waterproofed. There is a lot of advice available on the best way to do this while ensuring that you can remove the waterproofing easily should the need arise. Various forms of self-amalgamating tape, wrapped over with regular insulating tape, are commonly used and can be removed with a sharp knife. The author has a different approach, which is to wrap connectors with Blu Tack™, a recommendation that was given to me some years ago by a radio amateur who lived close to the sea and had found that this was about the only sealant that seemed to be oblivious to the ravages of a salt-laden atmosphere. Similar products (the normal purpose is for attaching posters to walls and similar) are no doubt available in other parts of the world.

REFERENCES

[1] WinSDR: http://psn.quake.net/WinSDR

[2] CW Skimmer: www.dxatlas.com/CwSkimmer

[3] Fun Cube Dongle: www.funcubedongle.com

[4] *VHF / UHF Handbook*, RSGB, p103.

[5] OZ2M transverter design: www.rudius.net/oz2m/70mhz/transverter.htm

[6] *VHF/UHF Handbook*, RSGB, p108.

[7] G3WOS 6m amplifier design: www.gare.co.uk/amplifier/index.htm

[8] Low Pass Filter: www.ham-radio.com/n6ca/50MHz/50appnotes/50tlpf.html

[9] *Computers in Amateur Radio* (2nd edition), edited by Steve White, G3ZVW, RSGB 2013.

[10] G4IFB on using Logger 32 and WSJT: https://tinyurl.com/y3akkgs7

[11] Grid Tracker: https://tagloomis.com/grid-tracker/

[12] JT Alert: https://hamapps.com/

[13] DXLite: http://g7vjr.org/dxlite

[14] Noise reduction video: https://youtu.be/qdQm17udCSo

[15] '[Not] Living with 6m/50MHz interference!': www.gare.co.uk/noise/index.htm

[16] ON4WW EMI-RFI page: www.on4ww.be/emi-rfi.html

[17] 'There's a Remote Possibility…', David Gould, G3UEG, *RadCom* August 2005 p80, September 2005 p76, October 2005 p76.

[18] *Remote Operating for Amateur Radio*, Steve Ford, WB8IMY, ARRL 2010.

Chapter 3

Antennas for 6m and 4m

YOUR INTEREST IN 6m or 4m may be to work local friends on FM, either simplex or via a repeater. For this you will usually use a vertical antenna, to give all-round coverage and because vertical polarisation is used for this purpose (primarily because it is easier to install a vertical antenna for mobile operation and also because, for local nets, it is important to have all-round coverage). For base station use you will probably choose a quarter- or half-wave vertical antenna, mounted at rooftop height or maybe higher (for example, if you already have a tower supporting antennas for other bands). There are many suitable commercial antennas available for this purpose. Some are multiband, so that you only have to install a single antenna to cover several of your favourite bands.

And in recent years it's become clear that a vertical (omnidirectional) antenna can be a boon for FT8 working too. During the 2020 Sporadic E season, for example, UK stations report working into the Middle East, the Caribbean and North America using just vertical antennas although a Yagi (see below) obviously helps to extend your capabilities, both in terms of distance and in terms of working the weakest stations. In any case, having an omnidirectional antenna as well as a Yagi can alert you to DX openings in directions other than that in which your Yagi is pointing.

But, ideally, for DX working, you will need a horizontally-polarised antenna with a useful amount of gain, though the strength of Sporadic E signals, as has been mentioned elsewhere, is often such that a very simple antenna, maybe just a dipole, will be more than adequate. However, for chasing those weak signals, a Yagi antenna of probably five elements or more is essential, mounted at the greatest height you can achieve in order to have a very low take-off angle (although W3LPL believes, based on experience, that anything above about 50ft can be counterproductive). If your previous experience has been on, say, 2m, then a 6m Yagi in particular will probably seem rather large, whereas if you are coming to 6m from HF it won't be so challenging, either to you, the neighbours or your other half!

A huge amount has been written about the merits and demerits of different antenna designs for 6m, as well as about the use of a single Yagi compared with a pair of Yagis stacked. Your choice may well be governed by practical issues, such as what it is possible for you to put on an existing tower in addition to antennas already *in situ*. Suffice to say that a poll of serious 6m DXers would show that the majority seem to favour a single long-boom Yagi of, perhaps, six or seven elements, or a stacked pair of Yagis. Where circumstances allow, a stack is probably the ideal solution, as discussed later in this chapter. In practice, relatively few amateurs are in a position to build one because their tower is likely already to be carrying other antennas such as an HF tribander and / or Yagis for the higher VHF bands. For 4m, given that most activity is within one hop of Sporadic E, a single Yagi of three elements or more is probably more than adequate.

Other designs of antenna can and do work on VHF, as you would expect – even wire antennas such as longwires or loops of various types. But, with the exception of a vertical antenna, in practice you will want to be able to rotate your antenna, which implies a more rigid design than can be achieved with wires. Some band users have successfully built quad antennas using garden canes and wire, but in practice some sort of aluminium construction is likely to be the order of the day. There are many good designs which you can build yourself and many more which are available commercially. The first consideration, though, is what your aspirations (and budget!) might be.

INNOVANTENNAS DESIGNS

One area in which there has been some interesting progress in recent years is Yagi design, with InnovAntennas **[1]** in particular developing some new and interesting designs. Justin Johnson, G0KSC, has developed three main types of new Yagi design, which he calls the LFA (Loop Fed Array), the OWL (Optimised Wideband Low impedance) and the OP-DES (Opposing Phase – Driven Element System).

The LFA Yagi has a rectangular shaped, full-wave loop driven element that is laid flat on the boom between, and in-line with, the parasitic elements. The smaller end sections of the loop driven element, which run parallel to the boom, are designed to be 180° out of phase with each other, therefore each side cancels the other out and so minimum radiation occurs from them. The main benefit is the rejection of unwanted noise, due to the highly suppressed side lobes combined with a particularly broadband design.

G0KSC's Optimised Wideband Low impedance ('OWL') Yagi is a low-

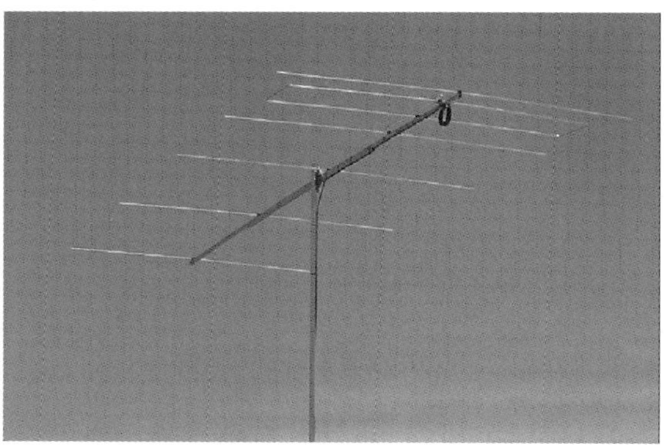

InnovAntennas 6m 6-element LFA Yagi.

InnovAntennas 4m 4-element 'OWL' Yagi.

7-element 6m OP-DES Yagi at the station of GM8IEM.

impedance (12.5Ω) design in which the split dipole driven element is changed to a folded dipole to transform the feed impedance to 50Ω. This allows for direct feed with 50Ω cable and removes any matching losses. While the use of a folded dipole in a Yagi is nothing new, it is generally done with 50Ω split-dipole designs to give a 200Ω feed impedance, which then requires a 4:1 balun or other matching device in order to be fed with 50Ω cable. G0KSC has computer-optimised the whole antenna with its folded dipole to provide a design with no matching losses and high front-to-back ratio.

The OP-DES Yagi has a 50Ω feedpoint impedance and all impedance transformation is done within the driven element itself and, therefore, the antenna is not susceptible to matching losses in the same way traditional Yagis are. The OP-DES has the ends of a split-dipole bent back towards the reflector element. This provides two functions. First of all, the current phase in each end section is in opposing phase and therefore cancel one another out; this helps to reduce / remove side lobes. Additionally, these end sections act as impedance controllers, taking this job away from D1 (first director) and thus making the adjustment of the antenna while in computer optimisation much easier, resulting in better levels of performance.

Justin has kindly developed two OP-DES Yagis specifically for readers of this book. Both are 5-element designs, one for 6m and one for 4m. Both are very wide band so, for example, the 70MHz design should be good for those Scandinavian allocations below 70MHz and the new allocations above 70.5MHz. The designs are shown in the accompanying diagrams (**Figs 3.1, 3.2** and **3.3**), photos and balun / feedpoint setup notes provided by Justin on pages 52 – 54. Justin also offers a number of free-to-use designs on his personal website **[2]**.

A 5-element 4m OP-DES antenna similar to those described here and installed at the G0JHC station.

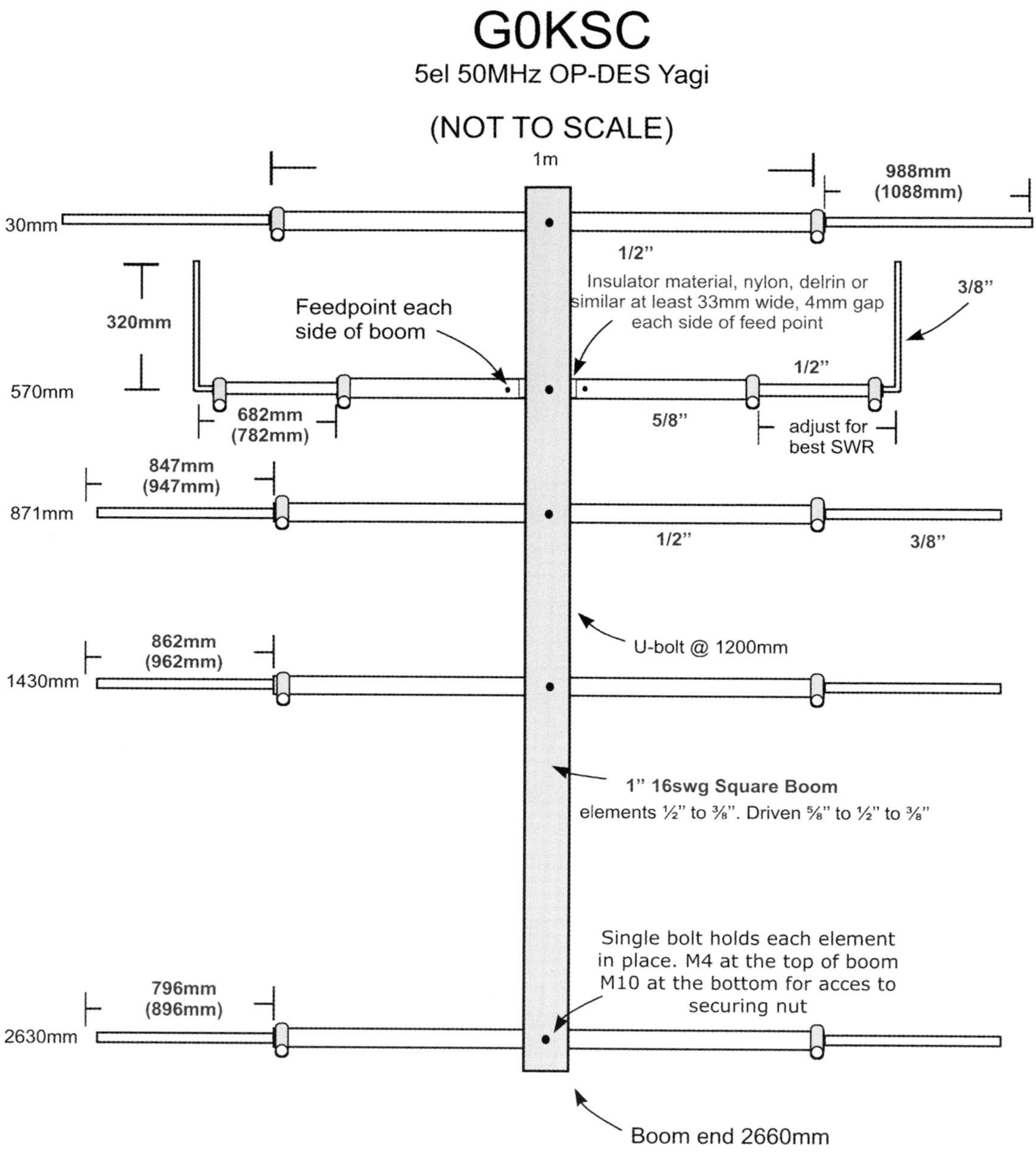

Notes:

Adjust DE width for best SWR
Install 1:1 choke balun at feed point
Boom Length : 2.660m
Feed system: 1 feedline with 1:1 balun or choke
D1 is shorter than D2
SWR bandwdith 49.95MHz to 50.6MHz better than 1.5:1
Boom will be 16swg wall. round tube should be 18swg to allow for each section to slide into the next

Fig 3.1: G0KSC 5-element 50MHz OP-DES Yagi design.

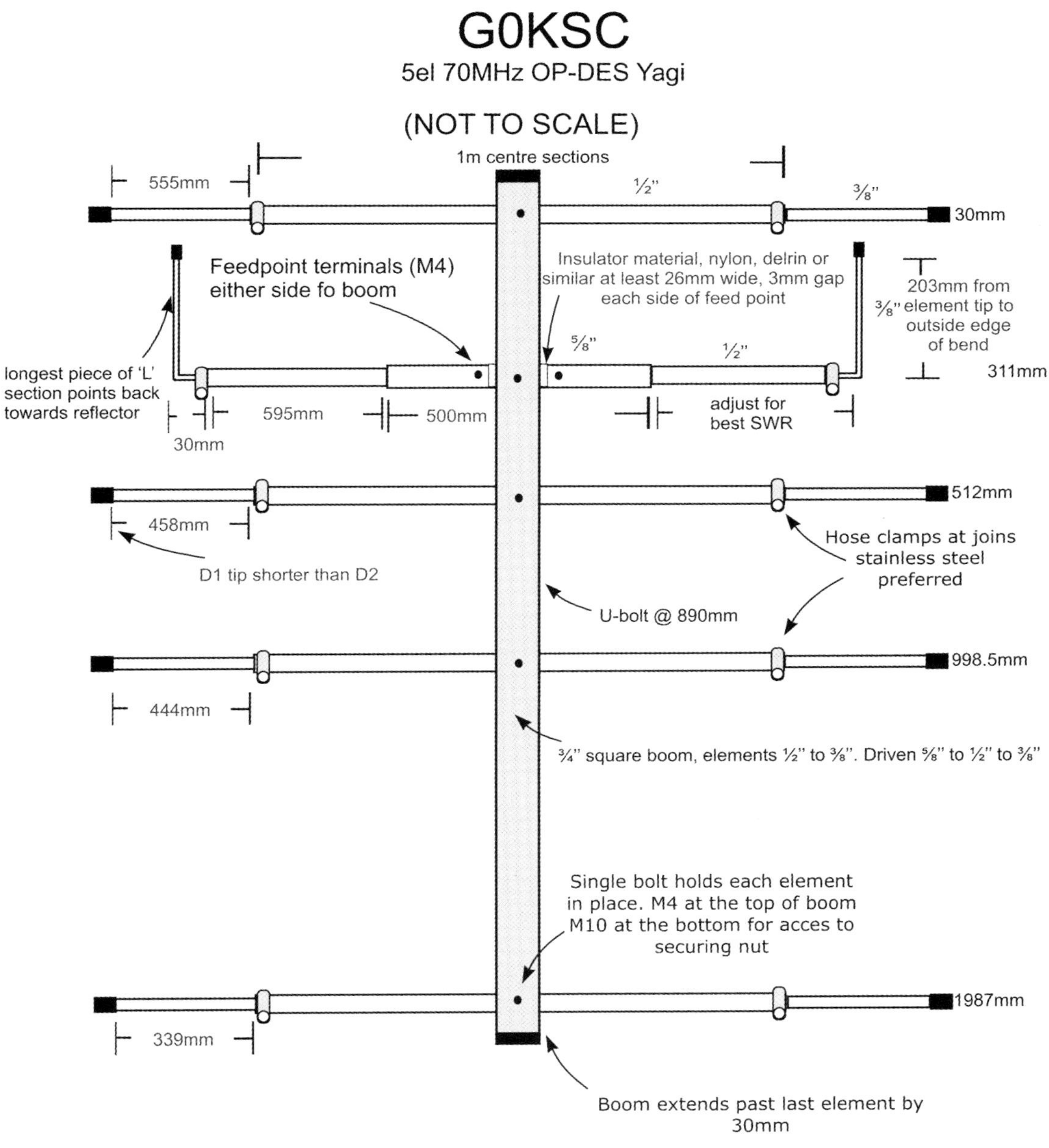

Notes:

Adjust DE width for best SWR
Install 1:1 choke balun at feed point
Direct 50Ohm feed, no matching device required
A common mode choke or current balun is required
SWR coverage 69.9MHz to 70.8MHz Better than 1.5:1

Boom will be 16swg wall. round tube should be 18swg to allow for each section to slide into the next

Fig 3.2: G0KSC 5-element 70MHz OP-DES Yagi design.

G0KSC DESIGN 6M AND 4M 5-ELEMENT OP-DES YAGIS

Feedpoint / Balun Setup Notes

The dipole centre should be bolted one side and pushed through the boom hole at the feedpoint, see photo below. When fitting coax to the antenna, better, more symmetrical, results will be seen by installing a choke or 1:1 balun at the feedpoint. Additionally, while good results can be seen by the coax / balun being fitted and running parallel to the boom, best results are seen if the balun / coax exits from the feed point at between 45° and 90°. In order to achieve this, loop the coax / balun with the insulators pointing towards the ground rather than to the sky. The baluns are fragile so make sure the weight of the coax cable feedline is not hung on the balun / balun terminals.

Secure your coax to the boom of the antenna first, prior to plugging into a fitted balun. **Fig 3.3** illustrates how the balun should look when connected to the feedpoint. While both are acceptable, absolute best results will be seen from ‘Option 1’. Once you have sufficient slack, fully tighten the coax supports on to the boom so as all of the supplying coax cable is fully supported by the boom with the small amount to slack to the connector. Now at this stage screw the balun connector to the coax connector and make a water-tight seal as appropriate.

You are now ready to play radio and make any finalisation adjustments. You may need to dip SWR to its lowest point.

Fitting the dipole.

Underside boom.

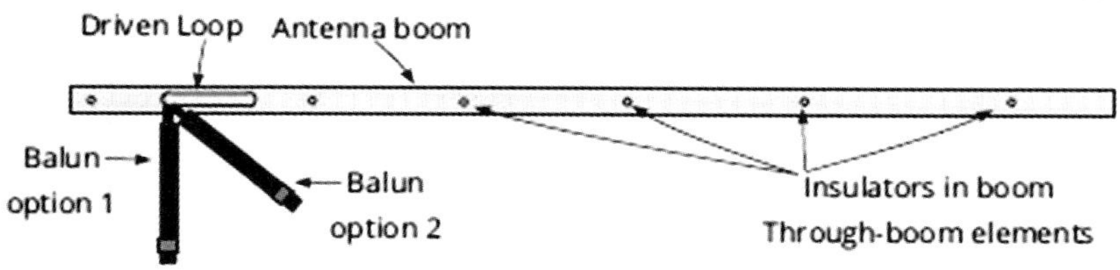

Fig 3.3: How the balun should look when connected to the feedpoint.

OTHER OPTIONS

There are, of course, many other suppliers of Yagi antennas for both 6m and 4m. The YU7EF website **[3]**, for example, carries designs for many different Yagis for both bands, depending on requirements. JK Antennas are very popular in the USA, while PowAbeam antennas, designed by Derek Hilleard, G4CQM, are sold in the UK by The DX Shop and are a popular choice. A Google search will throw up many sources of homebrew or ready-made antennas suitable for one or both bands. Chris Gare, G3WOS, has an excellent web page describing his quest for an improved Yagi for use particularly with EME **[4]**.

Indeed, up to a point, you can now select a Yagi to suit your particular situation, whether it's a location where you want to be able to null out particular local noise sources or a transmitter that doesn't like running into anything other than a very low SWR.

Of course, when choosing an antenna there are two factors to consider. One is the electrical performance, the other is physical design. Some expeditioners have gone to great lengths to develop Yagis which break down into short lengths and are (relatively) lightweight for their assembled size and performance. For home station use you would probably be much more concerned that the antenna be solid in its construction and able to withstand the worst your local weather can throw at it over a period of years.

WHAT AM I TRYING TO ACHIEVE?

There is an old adage in amateur radio which is that one can never have enough antennas. There is another, equally old, one which says that if your antenna didn't fall down last winter it clearly isn't big enough!

In practice, we all have to compromise. You may already have a tower and HF beam, in which case you will probably be thinking about mounting a VHF Yagi above it. If you want to be able to operate on both 6m and 4m, one perfectly practical solution is to have a duo-band Yagi: a number of manufacturers now offer several alternative designs, according to how large an antenna you can accommodate.

A 6m and 4m dual-band Yagi, with 4-elements on each band on a 2.1m-long boom.

Two of the 6m parasitic elements can be seen interlaced between the much longer HF elements on this SteppIR Yagi as it is being erected by Bert, PJ4KY, at his station on Bonaire.

Users of the popular range of SteppIR HF Yagis will also be aware that a set of parasitic elements is available for some models that creates a gain antenna for 50MHz. The pattern and gain are not on a par with a good monoband Yagi but it is certainly an easy way of getting started on the band if you have such an HF antenna. For many years I used a 3-element SteppIR that had parasitic elements for both 6m and 4m, although I believe the latter was a one-off, which was installed by the previous owner.

In contrast, for those who want to operate just on 6m and 4m, one effective solution adopted by serious DXers is to stack two 6m Yagis (as this is the band on which maximum gain is likely to be required), mounting a 4m Yagi between the two.

It is worth going back for a moment to talk about arrival angles and what that means for antenna height and design. Single-hop Sporadic E signals will almost certainly be arriving at relatively high angles whereas long-haul DX will require very low take-off and arrival angles. Thus, a low antenna may be adequate for the former, whereas a high antenna (with a long boom and multiple elements which tends to compress the vertical lobes) will be preferable for the latter. Stacking will bring down the vertical lobe even more. Height is less of an issue if, fortuitously, you have falling ground in the key directions. A take-off straight over the sea is even better, but few of us are fortunate to live that close to the coast (and, in any case, it will only help in that direction). In practice, the minimum requirement for a reasonably serious 6m DXer with no aspirations to EME is probably a 5- or 6- element long-boom Yagi.

That is not to say that DX won't be worked with something less – when the trans-Atlantic path is well-open on multi-hop or extended Sporadic E signals can, at times, be very loud indeed. But, equally, there will be expeditions and DX openings where the signals rarely climb much above the noise level and every dB counts. Yes, FT8 helps a lot and will retrieve signals that are actually below the noise level, but there is no getting away from the fact that the better the antenna, the more DX you will work!

Incidentally, when using stacked Yagis, it helps to be able to switch in either or both, in-phase and out-of-phase. This will help in selecting the optimum for the particular arrival angle of the signal you are trying to receive.

What the foregoing discussion makes clear is that there is no single solution as the choice will be determined by your operating interests, your existing installation, your home, garden and immediate surroundings and, no doubt, other factors too, such as planning (zoning) considerations and maybe even compromises to keep your own family happy.

SO, WHAT DO I DO?

Don't be put off. In reality most 6m or 4m operators will start with something quite modest, perhaps a 3- or 4-element Yagi for one or both bands, start to enjoy making contacts and then, human nature being what it is, think about something bigger and better for the next DX season. But it doesn't all have to be a case of 'suck it and see'. The good news is that there is plenty of software available nowadays for antenna modelling, not just of the antenna itself but of its surroundings, so that you can determine aspects such as the impact of other existing antennas (and hence how much clearance should be allowed when installing on the same mast). It really is worth investing a modest amount in such a modelling program before spending large amounts of money on an antenna which may not achieve what you want. Although most manufacturers nowadays quote antenna specifications in ways that can be compared (this hasn't always been the case), their figures will inevitably be presented in a way that shows their antenna in the best light and which will be based on modelling or measurements in the clear and certainly not take into account your local circumstances, with trees, adjacent buildings and whatever else.

For many years this author used his second tower for 160m in the winter and 6m in the summer, which worked reasonably well and might well do so for others. In that case, though, it would have been nice to have a second, smaller 6m antenna elsewhere in the garden to cover those 'out of season' 6m openings.

There is also an argument for a broadband (perhaps a log-periodic) antenna covering the 30 – 70MHz spectrum to feed your SDR or other general coverage receiver, in order to monitor frequencies outside the amateur bands themselves, while watching the MUF increase. Justin, G0KSC, of InnovAntennas has apparently been seeing increasing demand for broadband antennas for exactly this purpose in recent years, perhaps as SDR receivers become more commonplace. An alternative, if you want to try 4m as well as 6m, is a dual-band Yagi, such as that pictured on page 55.

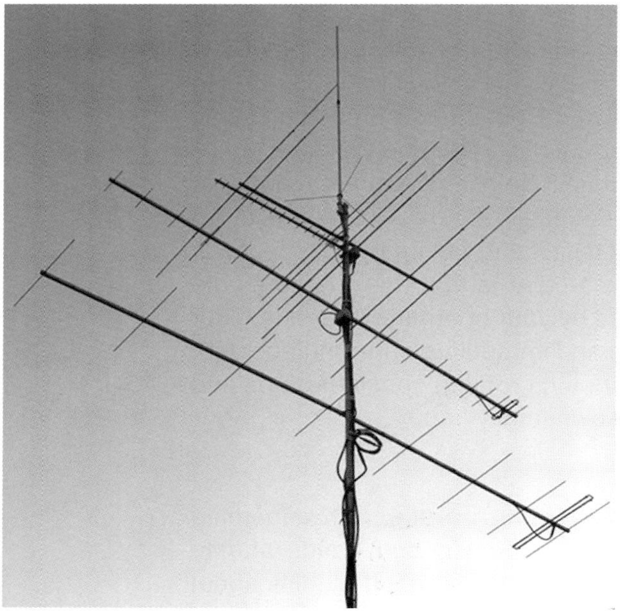

Antennas at G0KSC in 2020. The 8-element dual-band at the top uses a 1in square fibreglass boom and all elements are ½in centres with 3/8in tips, with the exception of the driven element which is 5/8in to ½in. A very light but strong arrangement.

The design of an antenna system should take into account what the antennas are required to do. In the case of 6m DXing, arrival angles of signals vary dramatically. Long-haul Sporadic E or F-layer signals can come in at (and need to be transmitted at) very low angles of just a few degrees above the horizontal. Short-hop Sporadic E signals can come in at angles of up to about 15°. No single antenna is going to give optimal performance in both situations.

Three possibilities present themselves. The first is to have something like a low dipole or short Yagi for close-in work and a second, high long-boom Yagi for the long-haul low-angle signals. This arrangement also has a potential advantage in spotting band openings because the long Yagi will have a very narrow beamwidth and may be pointing in the wrong direction as the band opens up. Equally, for contesting a narrow beamwidth can mean missing possible callers, so to have the two antennas and be able to switch easily and quickly between them could be a handy arrangement.

A second possibility is to have an elevation rotator similar to those used by satellite enthusiasts. But a 6m Yagi is much larger than most satellite antennas, so most 6m DXers will consider this option impractical. The third solution, much favoured by those with the wherewithal to do it, is to erect a stack of two Yagis. This is covered in some detail in the next section.

While few amateurs nowadays would want to build their own equipment, at least for serious DX and contest work, many people still prefer to build their own antennas, even for VHF. G0KSC's designs for both 6m and 4m 5-element OP-DES Yagis can be found on pages 52 – 54 and the G0KSC website **[2]** carries several other designs intended for personal use. Many other designs exist in the literature and on the Internet.

One problem, in the UK at least, used to be obtaining good quality aluminium in small quantities but there are now several retailers who offer this (search the Internet to find your nearest vendor), with tubing in suitable sizes for nesting and telescoping. Along with the various modelling software available, there really is nothing to stop the serious enthusiast from designing and building an antenna tailored exactly to his needs.

STACKING ANTENNAS

For those who are really serious about 6m DXing in particular (it is less important for 4m, as there is very little long-haul DX), the next step up is to stack antennas. Stacking helps to compress the vertical lobe of your antenna, reducing the take-off angle. By using a suitable switch-box you can also arrange to be able to listen and / or transmit on various combinations of the two antennas (both in-phase, both out-of-phase, top antenna only, bottom antenna only).

There are various ways in which antennas can be stacked. Kerry, G8VR, had a stack of three Yagis which, given his limited real-estate and taking account of other factors made sense for his location compared with, say, a two-stack of larger Yagis. Chris, G3WOS, for his EME array, has moved to a horizontal stack of two long-boom Yagis with elevation control (easier to achieve with the horizontal stack than with a vertical one) – see **Fig 3.4**.

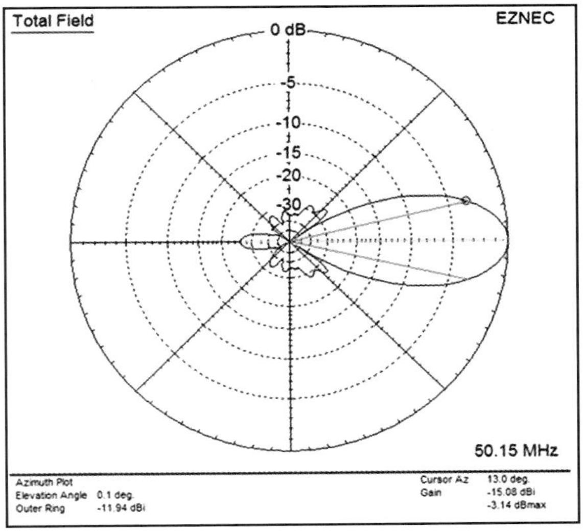

Fig 3.4: EZNEC plot of G3WOS's bayed pair of 7-element 6m LFA Yagis, showing the clean pattern with minimal side lobes.

A serious EME array, but only for those with deep pockets and plenty of space, will normally consist of a stacked and bayed array of four long-boom Yagis because this is really the minimum to be able to work a DX station who, at his end, has just a single Yagi.

Stacks have a number of benefits when properly designed and installed. Stacking increases the 'capture area' of your antenna system resulting in more gain for a given boomlength. A stack of comparable forward gain to that of a given single Yagi will have a shorter boomlength and wider beamwidth, which can be beneficial (the downside being that the stack is a three-dimensional system and therefore more of a challenge to erect).

Stack of 2 x 6-element 50MHz LFA Yagis at the station of VK6OX in Western Australia.

But the real benefits start to accrue by installing a switching system that allows you to select either Yagi independently or to use them together either in or out of phase, commonly

STACKING YAGIS

Life is simplest when both antennas are identical. The basic arrangement requires both antennas to be fed in-phase, with the power distributed equally between them. There are two ways of achieving this. One is to use a 2-way power splitter, such as those sold by RF Hamdesign **[5]**. Each antenna is then fed with an equal length of cable from the two output ports of the splitter.

The other is to use coaxial cable to make the feed arrangement. This can be done in various ways – it is not sufficient simply to have a 'Y' arrangement of identical types of coaxial cable as this will lead to impedance mismatches.

One solution is a completely symmetrical phasing harness using two equal lengths of 50Ω feeder, and then a quarter-wave transformer made from two paralleled lengths of 75Ω coax. The 50Ω cables can be whatever length is physically needed, so long as they are both the same. Then wherever you go in the band, the two Yagis will still be fed in the same phase. The paralleled 75Ω lines are also very easy to make up, taping the cables together and making the joints inside a small pill-box (after testing, fill the box with hot melt glue).

The difficulty is that they need to be an electrical quarter-wave long, which requires taking into account the velocity factor of the cable. Published data on velocity factors will only give you a starting point, so the lengths should be cut slightly over-long and then trimmed while checking the results with a suitable measurement tool (such as the miniVNA, AA-55 or AA-230 swept analysers – see 'Ancillary Equipment' section in Chapter 2).

The system works because the two 50Ω feeds to the antennas are in parallel, resulting in a 25Ω feedpoint. The paralleled 75Ω lengths have an impedance of 37.5Ω which, through a quarter-wave matching section, will result in an impedance of 56.25Ω where you connect the feeder back to the shack, a close enough match to 50Ω to be acceptable (mathematically, the sum is 37.5 x 37.5 ÷ 25). With this arrangement, and using antennas with a 50Ω feedpoint, there will be some mismatches in the system, but the power will be equally divided between the antennas and the SWR of the system as a whole should be acceptable.

There is a lot of useful information on the DK7ZB website **[6]**, follow the links on Stacking, where you will also find an alternative approach, also using paralleled 75Ω cables, which takes the same equation in reverse (in this case 37.5 x 37.5 ÷ 50), which says that the impedance on the far side of such an arrangement is 28.1Ω, so DK7ZB presents designs of 6m Yagis with a 28Ω feed impedance. This is fine if you wish to construct your own antennas, but most (all?) commercial 6m antennas are designed for 50Ω.

You can find information on stacking in a number of other places, for example in *QST* **[7]**, in *Six News* **[8]** and elsewhere, and on GM3SEK's webpage **[9]**.

called a 'BIP-BOP system' (Both-In-Phase, Both-Out-of-Phase). In phase they will have a low radiation angle, equivalent to a single antenna at a height of the midpoint between the Yagis. Out of phase they will have a high radiation angle. Until you have tried this capability, it is hard to understand just how useful it can be. The selection of either antenna independently has less of an obvious benefit but some users of stacks have noticed that when static levels are high the lower antenna may be receiving a much lower noise level than the top antenna of the stack and may therefore give better signal-to-noise on receive. Suitable switching systems are described in the popular antenna texts or ready-made systems can be bought from companies such as Array Solutions of the USA (some of whose products are distributed in the UK by Martin Lynch & Sons).

Whatever antenna system you decide on, you might want, as previously suggested, to model it in software at some stage, to get an idea of how it should be working and how it might be improved. Although the manufacturer will provide gain figures, in reality its performance will be affected by factors such as whether you have it stacked above an HF beam, how high above ground it is, the lie of the surrounding land, and so on. Nowadays all these factors can be modelled if you are so inclined.

Popular programs include Yagi Optimizer and EZNEC (both based on the NEC modelling algorithms) and terrain analysis software includes YT (bundled with *The ARRL Antenna Book* and designed primarily for HF antennas, but usable for 6m) and Terrain Analysis (TA). If you model single Yagis over real ground using TA or some type of professional modelling software such as K6STI's, you will find that they exhibit a deep null somewhere in their vertical pattern within the critical range of arrival angles from 4° to 15°. With stacked Yagis, where either top or bottom can be selected independently, one can choose to place the null at some point other than the optimum arrival angle for a particular signal. Using BIP / BOP / Top / Bottom switching it is almost assured that one of these positions will be optimal for more than 95% of all 6-metre signals you will hear, regardless of propagation.

ANTENNA SUPPORTS

There is nothing unique about antenna supports for VHF but a large 6m antenna and, in particular, a stack is by no means trivial and will require a substantial support. For a stack, the stub mast that supports the array needs to be heavy-walled tubing as the bending moment due to the upper antenna will be substantial.

There are many sources of information on masts, towers, winches and the other hardware required to get your antenna into the air. For example, I would particularly recommend GM3WOJ's website **[10]** for information about the care and maintenance of Versatowers, particularly popular in the UK. The TowerTalk reflector, hosted on www.contesting.com, is also a valuable source of advice and expertise.

THE FEEDER SYSTEM

Having decided on your antenna system, and a suitably robust rotator, do ensure that your budget will also extend to buying good quality feedline. It is no good spending money to get 1dB more gain out of the antenna, for example, while throwing away all that and more in feedline losses. A good rule of thumb is to aim for 1dB or less of feedline loss. Your choice of feedline will then be governed by what length of line you have between the shack and the antenna. **Table 3.1** shows the feedline loss for a number of popular types of coaxial cable at 50MHz. It now includes a number of products from Italian manufacturer Messi & Paoloni **[11]**. Their website also has a lot of good information about choosing and using coaxial cables. The list is by no means comprehensive but does give a good idea of what the options are and covers many of the coaxial cables likely to be used in amateur service.

It is worth noting that the old US military 'RG' specifications have been obsolete for many years, so all commercial cables with RG numbers are now 'generics' and may differ in detail from the original MIL specs. 'RG-8' was the first to be made obsolete, and has since spawned so many variants that the designation has become meaningless. In particular, the various kinds of 'Mini-8' bear no resemblance to the original MIL-spec cable, and the type and quality of construction also varies enormously – buyer beware! However, that cable is not normally recommended for use at 50MHz, except perhaps within the shack. The lowest specification of cable that can be recommended for long runs at 50MHz is 'RG-213' (that is the specification that replaced the old RG-8, and is itself now obsolete as well).

Below the line in the table for RG-213, there is a step decrease in attenuation per 100m. From here downwards, we're into the realm of 'serious' low-loss cables.

The next group are all cables of about the same outside diameter as RG-8 / 213 but with a larger centre conductor to reduce loss. This in turn requires a semi-airspaced or foamed dielectric to keep the impedance at 50Ω, but it is the larger centre conductor that's primarily responsible for the lower loss. Within this group, mechanical properties, shielding integrity and cost are the main reasons to choose one cable over another.

The cables towards the bottom of the table are much larger in diameter and therefore have lower losses. LDF5-50 is included because it's quite often available surplus at about the same price as LDF4-50 or even cheaper, largely because amateurs perceive it as being difficult to handle. In fact, it's hardly any different from 4-50 and connectors are cheaper too, so 5-50 can be a real bargain.

Data used in compiling the table came primarily from the Internet, with sources **[12]** and **[13]** being particularly useful and well worth a visit.

If your feedline length is such that the losses will be unacceptable, an alternative approach is to install a masthead preamplifier, as is common on the higher VHF and UHF bands. This results in greater system complexity, as you will need to power the pre-amp and switch it remotely from the shack,

Cable type	Loss @50MHz	Velocity factor	OD (mm)	Construction notes	Connector compatibility
RG-174	19.2	0.66	2.8	B, PE, T	I
UR76	12.0	0.66	5.0	B, PE. T	I
RG-58	11.0	0.66	5.0	B, PE, S/T	I
Mini-8	8.0 typ	0.75-0.80 typ	6.0 typ	B/BF, F, T	S1
Airborne 5	7.1		5.0	F	
Hyperflex 5	5.5		5.4	F	
UR67	4.6	0.66	10.3	B, PE, T	I
Aircell 7	4.5	0.83	7.3	PE	
RG-213	4.3	0.66	10.3	B, PE, T	I
Ultraflex 7	4.0		7.3	F	
Belden 9913	2.8	0.84	10.3	BF, S, S	S1
Ecoflex 10	2.8	0.86	10.2	BF, S, S	S1
H-100	2.8	0.84	9.8	BF, S, S	S1
Times Microwave LMR-400	2.8	0.85	10.5	BF, F, S	I
Ultraflex 10	2.7		10.3	F	
Hyperflex 10	2.7		10.3	F	
Westflex 103	2.7	0.85	10.3	BF, S, S	S1
Andrew LDF2-50	2.4	0.88	9.7	S, F, S	S2
Airborne 10	2.4		10.3	F	
Broad Pro 50/C	2.4		10.3	F	
Times Microwave LMR-500	2.3	0.86	12.7	BF, F, S	S
Ecoflex 15	2.0	0.86	14.6	BF, S, S	S2
Ultraflex 13	2.0		12.7	F	
Hyperflex 13	2.0		12.7	F	
Times Microwave LMR-600	1.8	0.87	15.0	BF, F, S	S
Andrew LDF4-50	1.6	0.89	16.0	S, F, S	S2
Times Microwave LMR-900	1.2	0.87	17.3	BF, F, S	S
Andrew LDF5-50	0.8	0.89	28.0	S, F, S	S2

Notes:
The top three are not recommended at 50MHz, except for patch cables and very short runs.
VK1OD's useful 'RF Transmission Line Loss Calculator' is at: http://vk1od.net/calc/tl/tllc.php

Construction notes:
Outer conductor: B = braided, BF = braid over foil, S = semi-rigid copper tube.
Dielectric: PE = solid polyethylene, S = semi-airspaced polyethylene extrusion, F = foamed (usually polyethylene).
Centre conductor: S = solid copper (copper-plated or copper tube), T = stranded copper (or copper-plated)

Connector compatibility:
I = industry standard for this type of cable, S = special, S1 = special: may fit a standard connector body but requires different braid clamp and has larger centre conductor, S2 = requires special connector or major adaptation of a standard connector.

Table 3.1: Matched attenuation of some 50Ω coaxial cables (in dB per 100 metres).

Peter, G8BCG, advocates the use of a mast-head preamp in an EME installation.

and it will inevitably be a potential point of failure. But a good quality preamplifier (and I stress good quality, because there is little point in having a transceiver with high dynamic range and intercept point if the preamplifier has neither) will get round those feedline problems. In practice most 6m DXers prefer to spend money on good quality feedline rather than a preamplifier, but a preamplifier is advantageous in certain low signal-level situations. That said, Lance, W7GJ, warns against using a masthead preamp in EME installations as he feels it is too likely to be a point of failure. Better, says Lance, to ensure less than 1dB feeder loss and, if necessary, have a pre-amp in the shack. In contrast, Peter Taylor, G8BCG, finds a preamp to be very helpful.

Chip Angle, N6CA, says in the application notes to his U310 preamp design, "When used in conjunction with good band-pass filters you can bring any good radio up to its best sensitivity without blasting it with 24dB gain from a GaAs FET preamplifier. Your radio won't handle that much gain anyway."

Designs for 6m preamplifiers can be found, for example, in the RSGB's *Radio Communication Handbook* and also on various websites (see particularly **[14]**, the N6CA design mentioned above). Commercial models are also available, for example the SSB-SP-6 from SSB Electronics or the LNK-50 from Hamtronics.

REFERENCES

[1] InnovAntennas: www.innovantennas.com
[2] G0KSC: www.g0ksc.co.uk
[3] YU7EF: www.yu7ef.com
[4] G3WOS 6m antenna: https://www.gare.co.uk/6m_antenna
[5] RF Hamdesign: www.rfhamdesign.com/products/powersplitters/2waysplitter/index.php
[6] DK7ZB home page: www.qsl.net/dk7zb
[7] QST, January 2008, p82.
[8] Six News, Issue 65, May 2000.
[9] GM3SEK on stacking: www.ifwtech.co.uk/g3sek/stacking/stacking2.htm
[10] GM3WOJ on lattice towers: www.qsl.net/gm3woj/latticetower.htm
[11] Messi & Paoloni: https://messi.it/en/home.htm
[12] GM3SEK website: www.ifwtech.co.uk/g3sek
[13] Cable loss data: www.qsl.net/dk3xt/cable.htm
[14] U310 preamp: www.ham-radio.com/n6ca/50MHz/50appnotes/U310.html

Chapter 4

6m and 4m Propagation

WHAT MAKES 6m unique among amateur bands is that it sits at the transition between HF and VHF propagation, so you get the best of both worlds. On the one hand, at the peak of the sunspot cycle you can enjoy superb F-layer propagation, with low absorption. This, coupled with the sort of high-gain antennas that are so much easier to put up for 6m than for the traditional HF bands because of their smaller size, means that it is possible to enjoy long-distance propagation that is at least as good as you may have experienced on 10m at the sunspot peak. Some very long-haul paths have been achieved this way, the UK to Hawaii path quoted in an earlier chapter being a good example. At the other extreme, serious 6m enthusiasts experiment with moonbounce (Earth-Moon-Earth or 'EME') which, while it works better on the higher bands of 70cm, 23cm etc is doable on 6m with enough power and a sufficiently sensitive low-noise receiving system. And what has transformed 6m EME has been the availability of the WSJT suite of software, which allows signals to be recovered from well below the noise floor of your receiver.

To get the most out of 6m, you need to familiarise yourself with the various propagation modes and take advantage of each and every one of them. East–West F2 propagation may only be possible for a few years around the peak of each sunspot cycle. Sporadic E (Es, or more correctly E_s) propagation occurs each year in most locations and can be relied on to produce DX contacts even at the bottom of the sunspot cycle, over quite considerable distances, especially in the case of multi-hop Sporadic E or when linked with other propagation modes such as Trans-Equatorial Propagation (TEP).

And it's important to remember, if you want to compete seriously in terms of countries and squares worked, that the major award programmes recognise contacts made by way of digital modes, provided they do not utilise active repeaters (whether ground or satellite based). So, a QSO via a repeater would not count for the 6m DXCC award, but a contact made via the moon or by bouncing your signal off the trail of a meteor certainly would. And for terrestrial propagation, FT8 has, in recent years, become the mode of choice for serious DXing.

The following sections describe the main modes of propagation you will encounter on both Six and Four (although F2 propagation is unlikely to be experienced on 4m, all other modes are equally relevant to both bands).

D64K was a surprise new one for many in the summer of 2012, worked from the UK almost certainly by a combination of TEP and Sporadic E.

But bear in mind also, when reading the following material, that quite often you may find yourself making QSOs using a combination of more than one propagation mode. For example, at some times of the year stations in Europe or the more northern parts of North America may be able to use Sporadic E to launch their signals into TEP which is happening well to the south of them. In many ways it is discovering how to 'plug in' to this sort of phenomenon that makes 6m so interesting compared with the more straightforward propagation modes you may use on other bands. A typical example was during the summer of 2012, when UK stations worked D64K and 3B8DB on 6m, when the solar numbers were modest, to say the least. The author worked FR4NT in May 2012 while running about 40W to a SteppIR HF antenna with 6m add-on. All these contacts were undoubtedly made through a combination of Sporadic E and TEP.

If you learn to recognise characteristics of the different modes you will be able to target your efforts in the directions and at the times most likely to bring results. This is easier to do on Six than on, say, 20m although sometimes the mechanisms at work can be quite involved. For example, one unusual contact was between Scotland and South America where the Scottish station was beaming northwards into an aurora. The backscatter signal from the aurora was then launching into what was probably Es propagation until it hit TEP across the equator!

But perhaps what is even more fascinating is that, even now, we only have a limited understanding of 6m propagation. A classic example is the way in which long-distance paths open up at certain times of the year, for example from Japan to Europe or to the USA, at times of sunspot minima, where the signals are solid, without flutter, for long periods in a way which cannot convincingly be explained by multi-hop Sporadic E. Attempts are being made to explain these phenomena, one mechanism proposed being 'SSSP' (Short Path Summer Solstice Propagation, see later in this chapter).

Right now, we cannot be certain of exactly what is happening and, in a way, that is the fun of the band. In 2020, during the Es season, new paths were discovered as a result of using FT8 – paths that presumably had existed previously but with so much attenuation that they were impossible using the 'traditional' modes (UK to Japan in the evening UK-time, for example). Although we might not know the exact causes of certain band openings, we can certainly enjoy the propagation that they bring.

The rest of this chapter covers the main modes of propagation with which we are concerned in broad terms. For more detailed explanations you will need to go to a specialist propagation textbook or dig out some of the more esoteric research papers that have been written. But it is worth saying once more that even the experts can be mystified at times by propagation on this band. The June 2006 opening between Alaska and Europe, a first for the band, came seemingly from nowhere at pretty much the bottom of the sunspot cycle and in 2007 and 2008, again at the bottom of the cycle, UK and US stations were able to work into Japan when there was no HF propagation at all over that path. Eternal vigilance seems to be the key. Expect the unexpected!

GROUND WAVE

While not a long-distance propagation mode, it is worth understanding the sort of distances that ought to be workable on 6m and 4m without relying on reflections and simply following a direct path. The starting point with radio propagation is that radio waves are just one component of the electromagnetic spectrum. Like other elements of that spectrum (visible light, X-rays), radio waves will travel in straight lines unless something acts to divert them from that path. That something could be a change in density of the medium (think of light waves being refracted as they enter or leave water), or a strong magnetic or electric field.

Even when travelling in a straight line, the waves will not go on for ever. They will be attenuated by the medium through which they are travelling. And they will disperse with distance, depending on the gain of the antenna from which they are propagated. Although a coherent beam of light from a laser, for example, can remain concentrated over extended distances, there is no straightforward way of achieving this with the wavelengths involved in radio frequencies. However high the gain of an antenna, the radiation will spread out with distance.

There is no simple answer as to how far VHF signals will travel, as it will very much depend on your location (hill, valley) and what lies between you and the station you are trying to work. Repeaters will be sited on hilltop locations to allow maximum line-of-sight distances but on both Six and

Four, as indeed on other bands, a certain amount of refraction is to be expected, along with possible diffraction over hilltops, to allow distances of several tens of kilometres to be achieved on a reliable basis. So even under flat conditions and without resorting to the more esoteric modes such as moonbounce, you should be able to work other stations in your locality when they are active provided that you don't live in the back of beyond! The better-equipped your station, in terms of power, antenna height and gain, and receive performance, the greater you will be able to extend your groundwave range. Bear in mind, though, that over a direct path polarisation will not change as it tends to when subjected to the vagaries of the ionosphere, so polarisation needs to be the same at both ends of the path. Repeaters use vertical polarisation because they are designed primarily to extend the range of mobile operators and it is easier to install a vertical antenna on a car than a horizontal one. So, if you want to work through a repeater or work mobiles directly you will need a vertical antenna of some sort. Fixed to fixed working tends to be on horizontal antennas, typically (but not restricted to) a Yagi, which will need to be rotatable.

And, as mentioned elsewhere, using FT8 will considerably extend your range under 'flat' conditions, often out to several hundred kilometres.

TROPO DUCTING

The troposphere is lower than the ionosphere and tropospheric ducting relies on the effects of weather on the atmosphere, specifically the existence of a temperature inversion. This temperature inversion can act, as far as radio waves are concerned, as the wall of a duct along which VHF waves are guided in the same way as in a metallic wave guide.

2m and 70cm operators are very familiar with this effect but it can also be used at 6m and 4m, although signals tend to be weaker and openings less frequent as the duct needs to be larger and larger ducts occur less often.

Tropospheric ducting is also more common in tropical areas and near large bodies of water. Sea paths of 2500km or so are not uncommon in parts of the world (such as Hawaii) where the necessary ingredients come together. The important thing is that the stations at both ends of the path must be able to launch signals into the duct to be able to communicate with each other.

SPORADIC E (Es OR 'E SKIP')

K1SIX, on his website, says "Regardless of solar conditions, Es remains the bread and butter of the average 6m DXer". Sporadic E propagation is caused by reflections from the lower levels of the ionosphere (90 – 130km), giving a normal range of between 700km and 2500km. Double-hop Es, to the eastern Mediterranean from the UK for instance, is quite common and triple- or quadruple-hop contacts to North America occur most years (a great example being G0LCS to VE7XF at 1807UTC on 10 June 2001).

At the other end of the scale, when the ionisation is particularly intense, ranges as short as 350 or 400km can be worked. As a rule of thumb, the more intense the ionisation, the higher the frequencies that will be able to be reflected, and it will become possible to work over shorter paths as the higher angle signals needed for those paths will be reflected back rather than passing straight through the E layer.

To take an example that would be typical from the UK, when Sporadic E is present over the centre of Europe and the MUF rises above 50MHz, contacts with Italy, Slovenia etc are likely, but in the first instance nothing will be heard from nearer countries such as Germany, because the ionisation will not be sufficiently intense to reflect the higher-angle signals that would be required to come back down at that shorter distance. But by the time the MUF has reached 144MHz, and 2m Sporadic E contacts with Italy become possible, it is likely that, from the UK, loud German, Polish and similar-distance stations will be workable on 6m. The Italian stations may still be audible – but may well have moved to 144MHz to enjoy the 2m opening.

Obviously, 4m falls between the two – as Sporadic E starts to affect 4m, the skip distance on 6m will have started to draw in.

Es conditions are literally 'sporadic' – they are unpredictable, developing quickly and disappearing again just as quickly. Signals can be (but aren't always) very strong but highly localised, frequently with rapid fading (QSB).

Although it can occur at any time of year, in the northern hemisphere Sporadic E is most likely between May and August, with a secondary peak

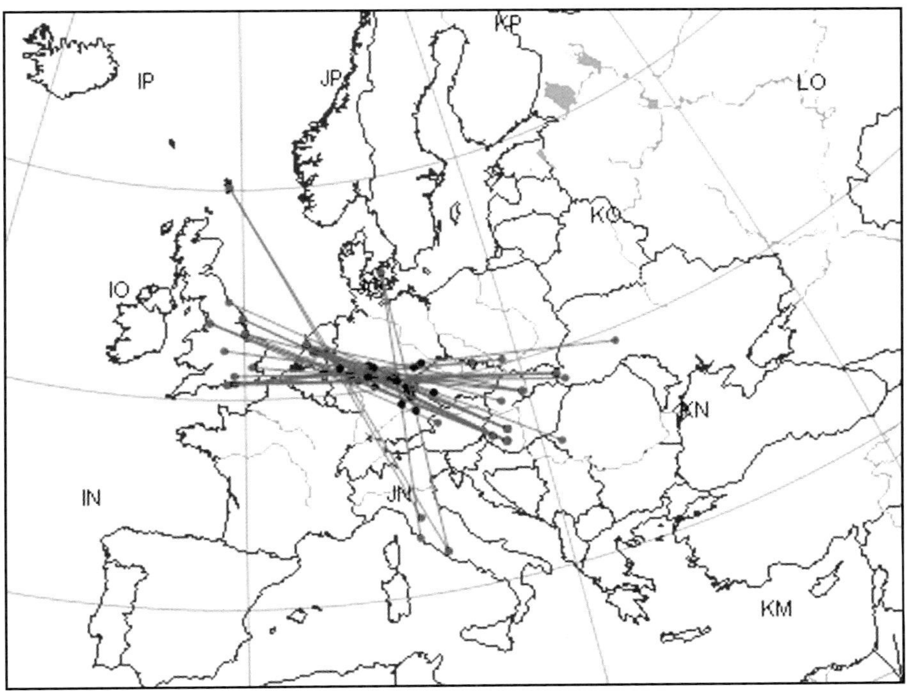

Fig 4.1: Plots of 6m Sporadic E QSOs made on 15 April 2006.

in December and January. Openings can also take place at any time of day, or even 24 hours a day, but the likelihood is higher mid-morning and around 1800 local time.

During a Sporadic E opening the centre of the 'cloud' will usually appear to move from south-east to north-west; this is not so much the cloud itself moving, but rather an effect of the rotation of the earth. **Fig 4.1** (from the G7RAU LiveMUF program) shows Sporadic E QSOs made on 15 April 2006, the first Es of the year as observed by Chris Deacon, G4IFX. Chris comments, "Because it was early in the year the Es was not particularly widespread and that makes the situation pretty clear, with a single Es cloud over central Germany."

Particularly for stations well away from the equator, Es accounts for the majority of international contacts during the years of solar minimum, although it's there at solar maximum too, some would say not quite as often. In the peak months of June and July, Es occurs somewhere in Europe and North America on most days. There is invariably a high level of activity as both 6m and 4m operators are very familiar with this form of propagation, and European operators will expect to be able to work many European countries as well as, on 6m, 'across the pond' to the Caribbean and North America. From the east coast of North America, most states will be workable, along with openings to Europe and North Africa and, on occasions, across the Pacific to Japan.

A high-gain, low take-off angle antenna can be a positive disadvantage when trying to work Sporadic E. The centre of ionisation can move around quite rapidly and arrival angles will typically be quite high. So, a small antenna, quite low, can turn out to be the most effective during Sporadic E openings. Perhaps this is another reason why this propagation is so popular, because a lot of DX can be worked with a modest station. This said, the longer-haul paths, launching into a Sporadic E 'cloud' right on the horizon, still demand an effective high-gain low-angle antenna most of the time, especially for double- or multiple-hop Sporadic E where signal strengths are likely to be much lower than with single-hop (every intermediate reflection from the earth, even if over the sea, incurs substantial attenuation).

Much has been written about Sporadic E and its causes and the matter still isn't resolved to everyone's satisfaction. Some argue that the ionisation results from wind shear, while others look elsewhere for an explanation. Some aspects are clear, such as the fact that Sporadic E has a diurnal pattern, peaks near the solstices, and occurs more frequently in latitudes closer to the equator. But none of the explanations that have been investigated have so far produced a 100% correlation or, perhaps more importantly, a means of predicting its occurrence.

Whatever its cause, the effect is to produce drifting 'clouds' of air that is ionised sufficiently to reflect radio signals. When the paths of QSOs are plotted after the event, there is often a clear centre to the reflections, showing that the ionisation is quite localised. But during the major periods of Sporadic E the ionisation can be quite widespread.

There is certainly plenty of literature to follow up if you wish to explore further. A good starting point would be the pair of articles that appeared in *QST* in 1997 **[1]**. One conclusion from the studies reported there is that Sporadic E is more common where the horizontal part of the earth's magnetic field is the largest. This would explain why Sporadic E is most intense in south-east Asia, but infrequent in South Africa, even in summer.

Some suggest that, like auroral propagation, Es has a habit of repeating after 27 days (one solar rotation). Whether or not this is true, there is certainly no harm in keeping a weather eye on the bands 27 days after a good period of Es.

A more recent set of articles worth locating is a series of three discussing Sporadic E on 50MHz, 'Extreme-Range 50MHz Es', that were written by KH6/K6MIO and W3ZZ. They appeared in *CQ VHF* magazine in the Fall 2011, Winter 2012 and Spring 2012 issues.

The best way to get a feel for Sporadic E propagation is to experience it for yourself because it really does have a unique 'feel' quite unlike other propagation mechanisms, with signals often changing from loud to inaudible (or vice versa) in a matter of seconds as the clouds of E-layer ionisation move around. But reports in the magazines also give an idea of what can be worked during such openings, a good example being the report by the late Gene Zimmerman, W3ZZ, in his September 2006 *QST* column of some excellent openings in June of that year between the USA and Japan.

Costas, SV1DH, also made a fascinating analysis of an extended multi-hop Es opening on 4 June 2006 across the Atlantic. **Fig 4.2** (from a DX Atlas screenshot) shows clusters of stations worked by SV1DH, with the most distant station (in EM50 square) being at 9600km. If the path is assumed to be multi-hop Es, the clusters fit very well to a hop distance of

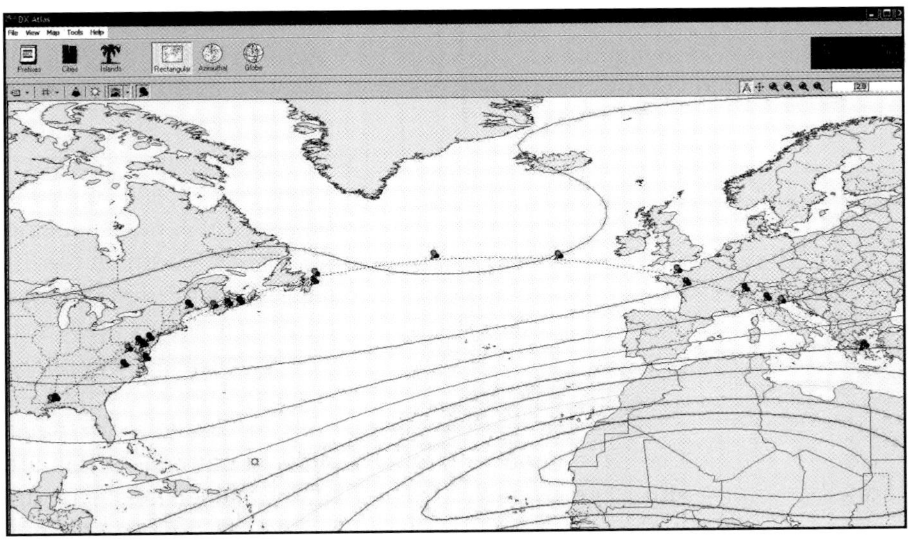

Fig 4.2: Stations worked by multi-hop Es on 4 June 2006 by SV1DH.

about 1200km, with heard / worked stations at six of the eight landing points, the other two terrestrial reflection points being mid-Atlantic where no stations would be expected. There are some who would argue that Sporadic E is such a variable phenomenon that the probability of getting so many Es clouds to line up in this way is vanishingly low, hence efforts to find other explanations (see SSSP, below), but others remain content that multi-hop Es is a perfectly rational explanation.

SHORT-PATH SUMMER SOLSTICE PROPAGATION (SSSP)

Over the past 15 years or so, 6m DXers have noticed that long-distance openings can occur around the summer solstice, even at times of low sunspot activity, with steady signals over a direct path between, say, Western Europe and Japan, or Eastern USA and Japan. This cannot be F2 propagation, as the F-layer MUF is far too low (indeed, from the UK, F2 propagation to Japan has rarely been observed over a direct path, it is almost invariably over a skew path). Multiple-hop Es is the generally-accepted mechanism, but the variable nature of Es propagation and the large number of intermediate hops that would be required (given that the E layer is considerably lower than the F layer) makes it hard to believe that the signals could be so steady and subject to so little attenuation en route.

Fig 4.3 shows a spectacular opening between Europe and Japan on the morning (UTC) of 14 June 2006, one of a remarkable sequence of JA–EU openings on successive days. A number of UK stations made QSOs with Japan. Was this by multiple-hop Es or by some new mode of propagation? Strong signals from Scandinavia were audible in the UK at the time and it can be seen from **Fig 4.3** that the path from the UK to Japan goes considerably further north than the path from central Europe.

Many 6m DXers have observed the phenomenon, which has been even more noticeable since the adoption of FT8, when stations are able to use the path even when signals are too weak for CW QSOs. Whatever it is, it may also explain some of the other paths that are observed at that time of the year which, again, can be stable over quite extended periods. Han Higasa, JE1BMJ, who is one of Japan's leading 6m DXers, with a well-equipped station that includes the ability to tilt his array and check arrival angles of signals, has postulated a mechanism which he calls Short Path Summer Solstice Propagation (SSSP). He first described his theories in the September 2006 issue of the Japanese

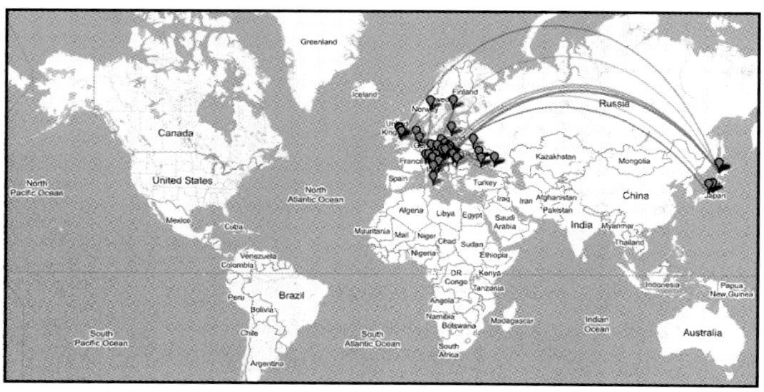

Fig 4.3: The spectacular 6m opening between Europe and Japan on 14 June 2006. (Map from dxers.info)

JE1BMJ's array which can be elevated as well as rotated, allowing a study of arrival angles.

magazine *CQ Ham Radio* and, with help from Chris Gare, G3WOS, they were translated into English for a multi-page article which appeared in the UKSMG's *Six News*. There isn't space here to do justice to Han's paper, but it is well worth a read **[2]**.

A later paper by Jim Kennedy, KH6/K6MIO, went into greater depth and argued that it is indeed possible to have multi-hop Es with signals propagating between Es clouds without intermediate ground reflections (chordal hop propagation). He gave several possible explanations for how this might come about but concluded that there is still plenty to be learned.

While SSSP cannot be predicted, it seems to be most likely when the K index is zero or 1 and the A index is in the lower single digits during the period two weeks before and after the summer solstice. Almost any geomagnetic activity raising the field into the unsettled range will eliminate it. It is also somewhat cyclic like ordinary Es, appearing for a few days and then disappearing for a few days. Perversely, SSSP may become less common as sunspot numbers (and hence geomagnetic activity) increase.

Whether or not you are convinced, Han has started a very useful debate and encouraged 6m DXers around the world to look out for these unusual openings and make contacts, thereby accumulating more data. One day we may know for certain exactly what is happening (and even be able to predict it!) In the meantime, we can continue to enjoy the benefits.

The 2020 season saw one night with JA and HL stations being copied and worked from the UK around 0300UTC, while on another night there was a very large JA opening during the 2200 to 0000UTC time period. It is very hard to reconcile these openings with 'traditional' multi-hop Es – even with the benefits of FT8 decoding, signals would not be sufficiently strong.

To conclude, it is interesting that the previous edition of this book had a quote from Lance, W7GJ, to the effect that new WSJT digital modes, by digging deeper for signals, might uncover new, hitherto unknown propagation opportunities. That is indeed what we have been seeing and, whatever the explanation, it's very exciting.

F2 PROPAGATION

HF operators are familiar with F2 as the mainstay of communications on the HF bands. But the Maximum Usable Frequency (MUF) at higher latitudes only reaches as high as 50MHz near solar maximum. It is caused by reflections (or, more strictly, refraction) at a much greater height in the ionosphere (300 – 400km) than Sporadic E, and therefore gives greater range: 3000 – 4000km for a single hop with multiple-hop taking signals out to several times that distance. As the name suggests, this propagation mode relies on refraction through the F2 layer of the ionosphere (the F layer effectively splits into two layers during daytime and recombines into a single layer at night – F2 is the higher of the two layers).

F2 propagation across the equator at 50MHz is not confined to sunspot maxima and is the basis for Trans-Equatorial Propagation (TEP), described later in this chapter. Stations at higher latitudes can often enjoy the benefits of TEP by launching into it via Sporadic E, one example of the sort of combined propagation mechanism that has already been mentioned. But this section is more concerned with east-west F2 propagation which is the basis for the really long-haul contacts that can be made on Six.

The best of the F2 action takes place around the equinoxes, during February – March and October – November. Signals can be strong but are more often quite weak. They normally vary much more slowly than Sporadic E and openings have a tendency to occur to the same region at the same time of day for several days running. Under good conditions the band can open shortly after daylight, firstly to the east, then following the sun's path across the earth's surface and closing to the west not long after dusk. If you are familiar with 10m propagation, then expect 6m to be similar in terms of the time of day and direction of the openings, but with shorter openings on Six than on Ten.

As a rule of thumb, east-west F2 propagation from higher latitudes (Europe, North America, for example) can be expected on 6m for only about three years around the peak of each 11-year solar cycle, though actual results will depend on the intensity of the cycle. In this context, Cycle 24 was very disappointing: at the time of writing predictions for Cycle 25 vary from very poor to very good – we can but hope! The 'SolarHam' web page **[3]** carries up-to-the-minute data. North-south F2 propagation is more frequent, benefiting from the higher MUFs in the equatorial zone (which are also responsible for TEP, see next section).

To work over the longest distances via F2, a low take-off angle (probably in the region of 2° to 5°) is recommended, so a long-boom Yagi at a good height (minimum one wavelength, but preferably somewhat more) and in the clear is essential for best results. Sloping ground in the target direction will help your take-off angle: an isolated hilltop location, preferably near the sea (saltwater in the near field also helps), is ideal! Stacking of antennas (covered in Chapter 3) can also bring down the take-off angle.

For a given take-off angle, the effective range of F2 propagation is greater than with E-layer propagation, as the F2 layer is considerably higher. **Fig 4.4** shows the sort of distances that can be expected for one- and two-hop propagation via the E and F2 layers for a range of take-off angles

The same considerations regarding F2 propagation apply to 6m as to the HF bands. For example, polar paths (signals travelling through the auroral oval) are going to suffer from significantly more absorption than east-west

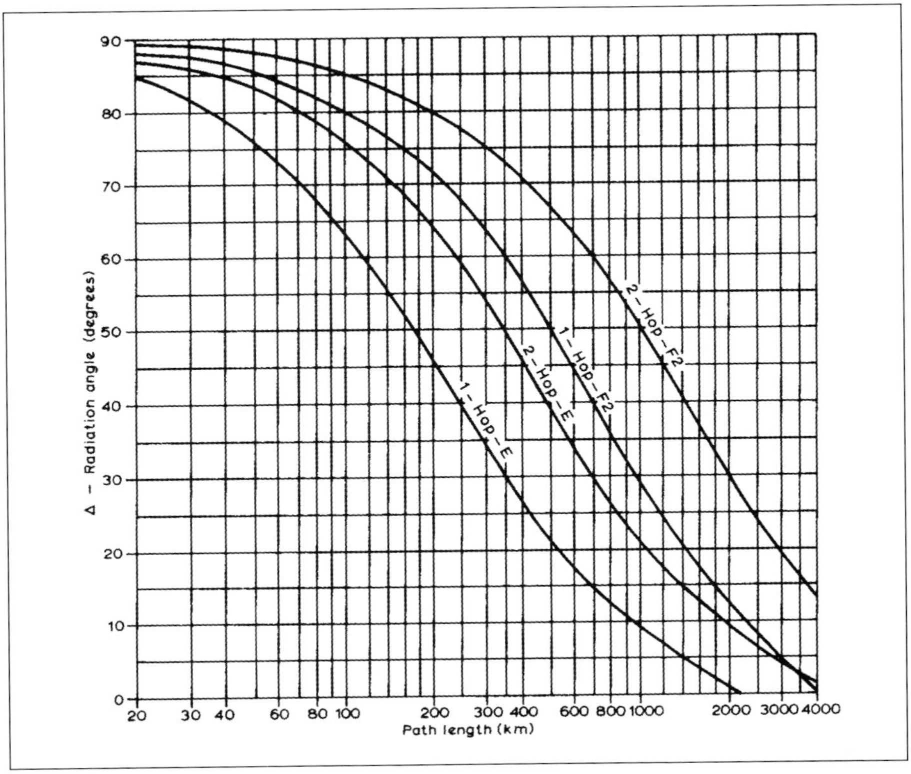

Fig 4.4: The normal distances of one- and two-hop E and F2 layer propagation for a range of take-off angles.

and north-south paths, especially when auroral activity (as indicated by the A and K indices) is high.

This is not the place to go into great detail about how F-layer ionospheric propagation works but, essentially, ionospheric propagation depends on the build-up of ions high in the fringes of the atmosphere where wave and particle emissions from the sun arrive with sufficient intensity to ionise the air (in other words to split the molecules into positively and negatively charged ions). This layer of ionised air then behaves almost like a mirror as far as radio waves are concerned, causing them to be returned to earth. However, the higher the frequency of those radio waves, the higher their energy, and at any given time there will be a frequency above which all radio waves will have sufficient energy to pass through the ionosphere into space.

That frequency will vary according to the level of ionisation. One measure is the Critical Frequency, the highest frequency at which signals which are transmitted *vertically* will reflect to earth. This will be much lower than the MUF (the maximum usable frequency for signals launched towards the horizon). Thus a Critical Frequency of around 15MHz would probably equate to an MUF of around 50MHz.

The actual level of ionisation will vary according to the amount of energy being received from the sun. And the energy received from the sun varies constantly. There are several cycles involved here: there is the daily effect due to the earth's rotation, there are seasonal effects, just as with the weather, and there is the 11-year sunspot cycle (more accurately, a 22-year cycle, as the polarity reverses every 11 years) superimposed on all those. Disturbances on the sun can also adversely affect ionospheric propagation and these often repeat on a 27-day basis, this being how long it takes for the sun itself to rotate with respect to the earth and present its same face back to us again. Many textbooks explain these effects in great detail as ionospheric propagation has been well studied over the years.

One of the benefits is that there are now many software tools for predicting ionospheric propagation, though their utility for predicting 6m propagation is limited. Any east-west F2 propagation on 6m is likely to be on the fringes of the MUF, dependent on particularly dense patches of ionisation (the ionosphere is never homogeneous) so over a given path the MUF may quite literally be drifting above and below 50MHz in a way that causes signals to appear and disappear in a matter of minutes, or even seconds. Any prediction tools can, at best, simply give an idea of when the MUF on a given path is likely to rise above, say, 30MHz such that it may be worth keeping a weather eye on Six, or when the MUF is somewhat lower and F2 propagation on Six is unlikely in the extreme. But remember that most propagation software was developed originally for commercial use and therefore errs on the side of caution (commercial users are looking for a reasonably reliable path) whereas we as amateurs are happy to settle for the most fleeting of band openings. Also, most such software assumes a typical HF station of, say, 100 watts and a 3-element Yagi, whereas serious 6m operators will have substantially higher ERP (effective radiated power) at their disposal. The trick is to identify those paths where there is a possibility of an opening and then to sit patiently at the radio listening for beacons, DX activity or, perhaps, monitoring commercial frequencies below 50MHz to follow the MUF as it rises. These techniques are described in more detail in the operating chapters of this book.

Solar data (used as the basis for HF propagation predictions) is readily available from a number of sources, giving an indication of solar flux as well as any likely geomagnetic disturbances (the A and K indices) which could adversely affect long-distance propagation, especially through the polar regions. As a rule of thumb, east-west propagation on Six from European and North American latitudes tends to coincide with those periods of the sunspot cycle where the Smoothed Sunspot Number is over 100, or if the solar flux is over 200 for two or more days with a little geomagnetic activity (K = 2 or perhaps 3). But your mileage may vary!

Anyone familiar with 10m propagation knows that some of the tougher paths (e.g. from the UK into the Pacific) can often be worked more easily via long-path propagation than short-path, as the short path would take signals though the northern auroral oval, with significant attenuation. The same is true of Six. High ERP is necessary to exploit these paths, along with low take-off angles and a good (high-gain) receiving antenna. The good news is that 6m antennas are significantly smaller than HF antennas and can be mounted at greater heights in terms of multiples of a wavelength.

Stations near the equator enjoy some excellent long-path propagation on 6m at times of sunspot maxima. Costas, SV1DH, reports long-path openings to Japan (30,900km!), with weak signals, but occurring earlier in the sunspot cycle than short-path propagation was noted. Also to Hawaii (26,600km), with no short-path openings observed to KH6. This KH6 path was also observed on a regular basis from Malta during the same period of Cycle 22.

Despite some of the excellent results which have been achieved on Six via F2, it is interesting to note that no G – ZL contact has yet resulted, though this path is relatively common on the 10m band at solar maximum. G – ZL contacts have had to rely on moonbounce.

TRANS-EQUATORIAL PROPAGATION (TEP)

Trans-Equatorial Propagation (TEP) is a variant of F2 which occurs mostly during local afternoon because of two zones of maximum F-layer ionisation which are located just north and south of the geomagnetic equator. **Fig 4.5** is a map of the geomagnetic equator and shows how this follows a somewhat different path to the actual equator, simply because the geomagnetic north and south poles do not coincide with the geographic north and south poles.

'True' TEP is observed when signals are reflected by both of these zones, possibly without a ground-reflection in between. The stations involved must be conjugate, i.e. roughly equal distances either side of the magnetic dip

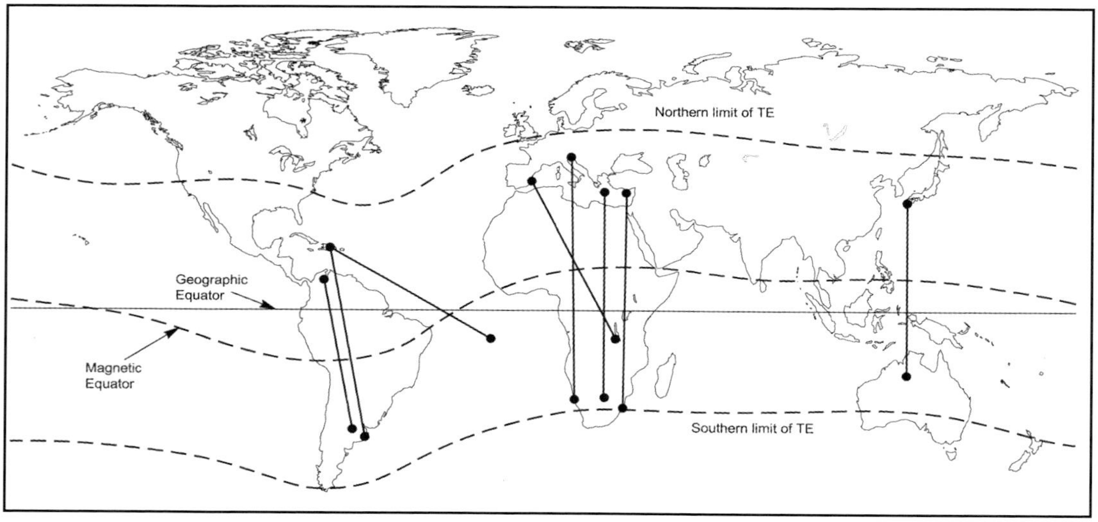

Fig 4.5: Map showing how the geomagnetic equator follows a different path to the actual equator (image courtesy of *ARRL Handbook*).

equator, which means that it is only possible to work southern Africa, Indian Ocean or South America by TEP from Europe.

TEP is most common between February and May, and between August and November, often with Es assistance, but can occur at any time of the year, particularly at lower latitudes (closer to the equator). TEP continues even away from sunspot maximum, although it does tend to cover closer distances or disappear at sunspot minimum.

Because the incidence of F2 drops off rapidly the further north you go, stations in areas such as the Mediterranean and the Caribbean tend to have a better time of it than those of us in places such as the UK or the northern states of the USA. But, as mentioned previously, quite often mixed-mode propagation, with a single Es hop, for example from the UK down to lower latitudes linking up to TEP from there on, can allow more northerly stations to access the DX especially to Africa and South America, in which case the openings can last well into the evening.

During Cycle 21 Euro-African sector TEP was studied extensively by a dedicated group comprising Ray Cracknell, ZE2JV (G2AHU), and Fred Anderson, ZS6PW, in Africa and Costas Fimerelis, SV1DH, in Athens, taking propagation time delay measurements and observing that it comes in two different modes, the normal 'afternoon mode' (A-TEP) lasting up to 1700UTC and supporting frequencies up to 70MHz with strong, steady and clean signals, and the 'evening mode' (E-TEP) between about 1730 and 2200UTC characterised by weaker propagation and heavily distorted signals up to 432MHz propagating through an ion depletion layer above F2 daylight heights. Propagation drops out completely between the end of one mode and the onset of the other. These times would obviously need to be adjusted for other longitudes, such as the Americas. ZE2JV, ZS6PW and SV1DH

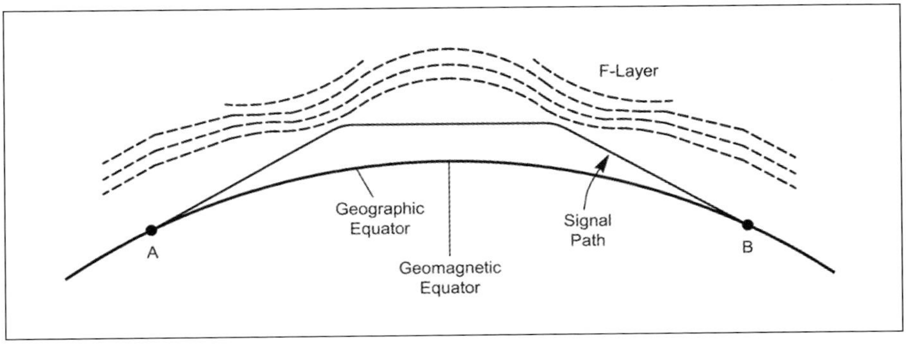

Fig 4.6: Typical cross-section of a trans-equatorial signal path (image courtesy of *ARRL Handbook*).

described their findings in the November and December 1981 issues of *QST* and although the studies were conducted at 144MHz the articles are well worth a read: this is a case where amateurs investigated a propagation phenomenon that was largely unknown in the professional world and the studies formed the basis of Costas's PhD. An extensive bibliography of 144MHz TEP appears at **[4]**.

Fig 4.6 shows a typical cross-section of a trans-equatorial signal path, showing the effects of ionospheric bulging and a double refraction above the normal MUF.

BACKSCATTER AND SIDESCATTER

Backscatter and Sidescatter are also interesting and useful modes. They occur when a small portion of the signal is reflected, from the Es or F2 ionisation itself or from the ground, either back towards the transmitter (backscatter) or to the side (sidescatter). Particularly strong scatter can happen when the signal hits a rough sea. Scatter signals have a hollow, echoing sound which is quite distinctive.

Backscatter can give QSOs with stations a few hundred kilometres from you and can be a useful way of filling in missing grid squares, as well as giving a strong indicator of a possible 'normal' opening in the direction from which the scatter comes.

But it's sidescatter that really comes up with the goods. The vast majority of the contacts which have been made between the UK and Japan via F2 (as against multi-hop Es) have been by sidescatter from a point somewhere in (or over) the Indian Ocean. The lesson of this, once again, is to make sure that you listen carefully – the best beam heading for those contacts with JA was 90° – 100° rather than the direct path at 30° – 40°. During the fabulous F2 conditions accompanying the secondary peak of Cycle 23, on many days stations in the US worked southern and central Europe direct path at 45° while simultaneously working northern Europe, including Scandinavia and stations as far north as JW and JX, via a skewed path at 140° – 145°. The skewed path signals were very strong, often over S9.

The concept is familiar to HF operators who often find at periods of low sunspots that trans-Atlantic contacts can be made by beaming to the equator somewhere midway along the path. During major contests, when many well-equipped stations are active, this method is particularly effective. On 6m similar considerations apply – signals will be weak but this can be countered by using a high-gain antenna and as much power as you are permitted.

AURORAL PROPAGATION

The natural phenomenon known in Europe as the Northern Lights has been studied and used as a propagation medium by amateurs for more than 60 years. Like the other propagation modes already discussed, auroral propagation is a result of the sun's energy impacting the atmosphere, but in this case it is the effect of a solar flare emitting bursts of energetic charged particles which stream outwards from the sun and spiral towards the earth via the solar wind. These particles are divided by the earth's magnetic field and then follow the field lines to regions known as the auroral zones. These zones are oval shaped and extend outwards from the poles to a radius of 23° on the night side of the earth and to 15° on the daylight side.

Visual auroral sightings indicate where the charged particles impinge on the earth's upper atmosphere, ionising the E layer at a height of 110km. The number of auroral openings in any year is dependent on the solar activity. Some areas of the sun can remain active for several weeks, causing repeats of events 26 to 28 days after the initial aurora. This is due to the period of rotation of the sun and events like these are known as solar repeats.

Aurora at Kangerlussuaq, Greenland, in September 2020.

Many such auroral events are able to support the reflection of 6m and occasionally 4m signals. It is usually quite simple to recognise signals which have been reflected from an auroral zone. They have a rasping sound and the point of reflection will normally be well to the north so that, instead of beaming directly at the station you want to work, you beam at the centre of auroral activity.

Warning Signs of Impending Auroral Events

Large increases in the solar noise levels on 95, 136 and 225MHz can be measured during flares, these frequencies being those used by professional observers. However, these increases are also very noticeable on 6m. These sudden ionospheric disturbances are often followed by short-wave fade-outs on the 14 to 28MHz bands, causing disruption of amateur and commercial HF communications. This is caused by cosmic particles and X-rays ionising the D-layer which absorbs, rather than reflects, radio signals. These particles complete the 150-million kilometre journey from the flare region on the sun in less than 15 minutes. It should be noted that short-wave fade-outs occur which are not accompanied by auroral events, but fade-outs followed by large magnetic disturbances almost always signal an impending aurora.

A good and reliable indication of a forthcoming auroral event is known as *pre-auroral enhancement* and is familiar to HF band operators. This effect is particularly obvious during periods of poor propagation such as in the summer months at sunspot minimum. A good example occurred on 6 April 1995 when, during a sustained period of mediocre conditions on 14MHz, the band became full of very strong signals from Australia and New Zealand with the propagation akin to that during sunspot cycle peaks. The following day HF band conditions had collapsed and there was a very intense aurora enjoyed by VHF operators on the 50, 144 and even 430MHz bands.

Shortly after the commencement of a magnetic storm the comparatively slow-moving particles ionise the E layer and align along the earth's field lines. VHF radio signals beamed towards the auroral regions are reflected and refracted by the moving area of auroral ionisation. This moving reflector causes frequency shift and spreading, making all auroral signals sound distorted and difficult to copy. Morse signals are transformed into a rough hissing note

The 2012 edition of Charlie Newton's, G2FKZ, classic book, *Radio Auroras*.

and SSB voice transmissions vary from a growl to a whisper. The amount of frequency change on signals varies proportionately to the frequency band used. This effect is known as Doppler shift and can be as much as 1.5 to 2kHz HF or LF of the actual transmit frequency on 144MHz. However, the frequency change is less on 6m, making auroral contacts on Six somewhat easier. If the aurora is strong enough, a form of Sporadic E develops, known as Auroral E (see below).

To learn more about what causes radio auroras, you can do no better than turning to *Radio Auroras* by the late Charlie Newton, G2FKZ. It was reprinted by the RSGB in 2012 and updated with new material by Neil Carr, G0JHC **[5]**.

Practical Considerations

From the point of view of the 6m or 4m operator, once it is clear that an auroral event is underway (black-out on the HF bands, auroral spots on the PacketCluster, information from other sources), it is necessary to beam at the main auroral 'cloud'. This will be broadly to the north but not directly north, so you will need to try various beam headings until signals peak.

It is important to be aware that auroral propagation is a scatter mode. Signals are not refracted as with F-layer propagation but, literally, scattered in all directions from regions of intense ionisation. As a result of this scattering signals can be heavily distorted and CW is generally much easier to copy than SSB under these conditions. As mentioned, auroral signals have a characteristic rasping sound quite unlike anything you will experience as a result of other propagation modes, so there can be no doubt of the propagation mechanism involved. **Fig 4.7** illustrates this quite nicely.

Auroral backscatter has two well-defined daily peaks, the first between about 1600 and 2000 local time and the second, during major magnetic storms, around midnight local time. During these events distances up to

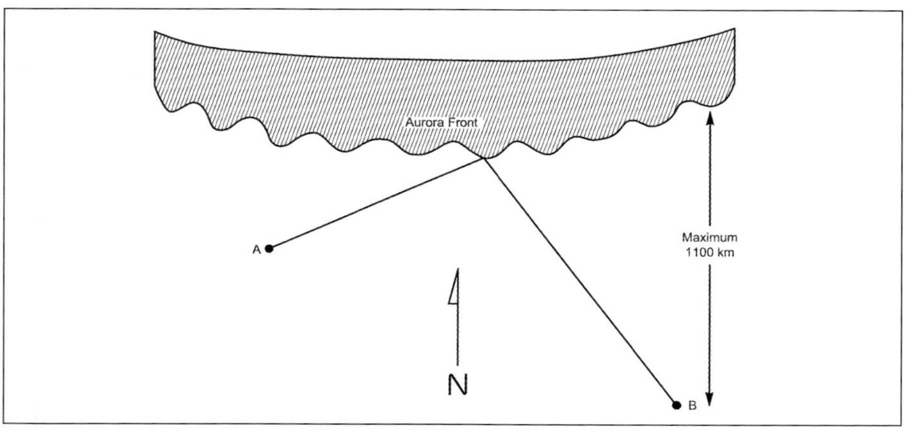

Fig 4.7: Auroral propagation. Point antennas generally north; optimal antenna headings may shift considerably to the east or west depending on the location of the aurora (image courtesy of *ARRL Handbook*).

about 2200km are likely to be workable. The strongest auroras can transform into a condition known as Auroral E, in which distortion levels are lower and path lengths even greater.

It is important to note that while what is generally referred to as 'auroral propagation' is a backscatter mode, Auroral E is *forward* scatter. The characteristics of the two are therefore rather different: auroral propagation has a rasping sound and the range is generally quite limited; Auroral E produces much cleaner signals and longer ranges. Thus it is not uncommon, for example, for UK stations to hear Doppler-shifted auroral signals from southern Norway and, at the same time, to be hearing clean Auroral E signals from northern Norway.

Predicting Auroral Events

Before the advent of the packet radio network most operators relied on the telephone to alert others about auroral openings. These links were quite successful.

Nowadays serious DXers monitor the DX Cluster and the Internet warning systems which provide a real-time record of events as they happen.

Magnetometers are live on the Internet showing you what is, and what will be, happening, and full daily solar geomagnetic reports are also to be found on the Internet. Many operators keep special 27-day auroral calendars on which they record both the visual and radio events which are reported in *RadCom* and on the GB2RS news bulletins. The auroral calendar in **Fig 4.8** which started on 1 January 1977 and records auroral events until 1 August 1978 shows the correlation between the visual and radio events and demonstrates the fact that auroras often repeat 27 days later. A study of **Fig 4.8** shows that in the period 1 August 1977 to 1 August 1978 UK observers

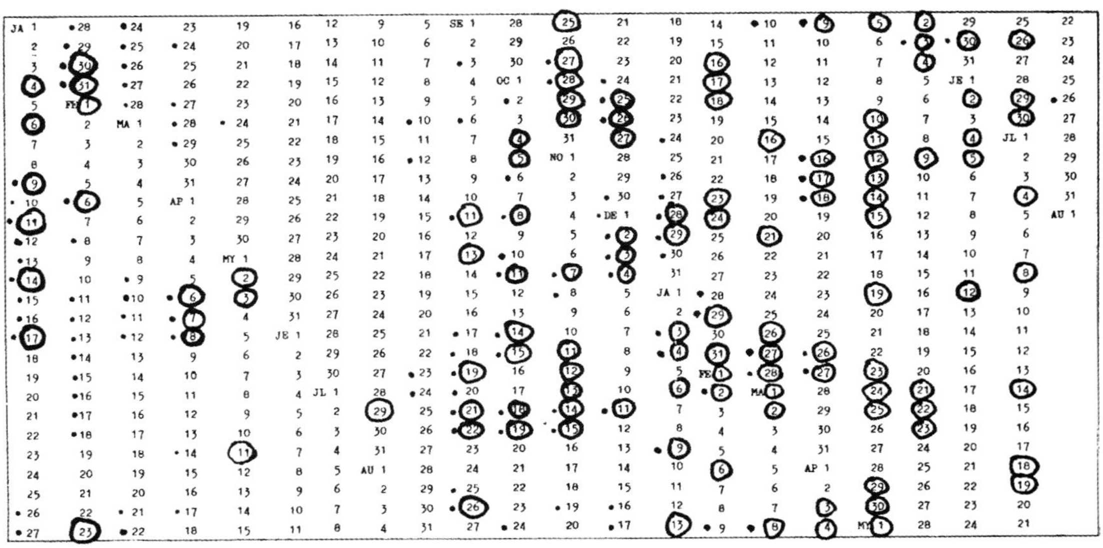

Fig 4.8: A 27-day auroral calendar.

recorded 70 visual aurora and 100 radio events. These numbers are considerably higher than during the same period for the previous two years, thus reflecting the increase in solar activity as Cycle 21 got under way.

The auroral warning calendar is very easy to use – simply circle any radio events and dot any visual auroras. An operator who was on for the 21 September 1977 event would have been prepared for the repeats which occurred on 18 October, 14 November and 11 December 1977. As the calendar shows, many events repeat in 26 to 28 days, often three or four times, proving that this method can be used successfully. It is also interesting to look at the calendar with hindsight – with the exception of 8 December 1977, the aurora of 18 September repeated consecutively 27 days later seven times. Did we miss an aurora on 8 December or was the event of 4 January and its subsequent repeats unconnected?

Beyond Internet alerting nets, from a practical standpoint one can predict radio auroras to a great extent by looking at certain types of geomagnetic data and comparing these to the time of day these events occur. Thus it is necessary to have both propitious geomagnetic circumstances and favourable geometry to produce effective radio auroras. The geomagnetic values are a K index of 4 or more (www.swpc.noaa.gov/rt_plots/kp_3d.html for the last three days planetary K indices) and a southward (negative) IMF (interplanetary magnetic field) Bz component (from www.swpc.noaa.gov/ace/MAG_SWEPAM_6h.html for the last six hours). In addition, the geometry has to be correct. The particles must be travelling earthwards and the geomagnetic storm has to strike at the proper moment during the day, preferably during and early into the afternoon aurora period commencing at 1600 local time. If any of these parameters fails to occur: a negative Bz, a higher than 4 K index, an earthward facing particle stream hitting us mid to late afternoon, the result is likely to be a poor or non-existent radio aurora.

The fact that amateurs note, record and make radio contacts via the aurora is of great interest to professional scientists studying the auroral phenomenon. Predicting correctly the date of an aurora and enjoying the benefits by way

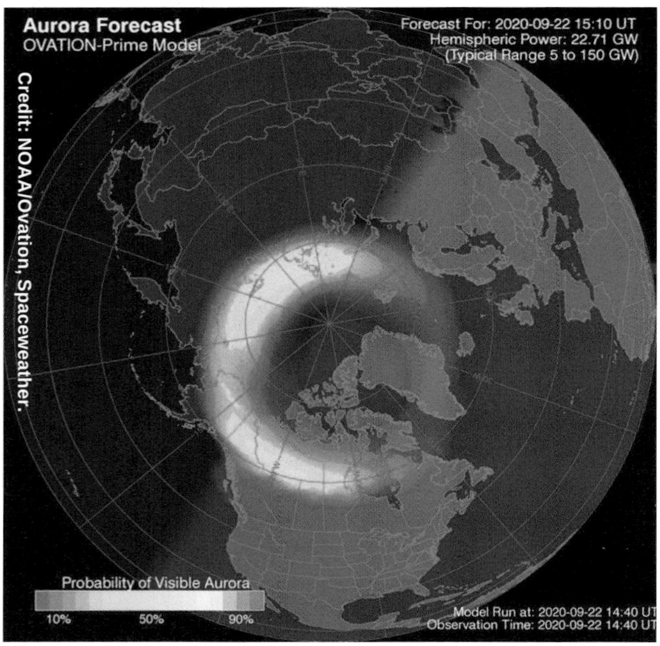

Aurora forecast for 22 September 2020, showing the auroral oval, from the Spaceweather website.

of DX is very satisfying. The fact that no two auroral events are ever the same and that it is impossible to predict which countries will be worked simply adds to the attraction of working DX via the aurora.

METEOR SCATTER PROPAGATION

Meteor scatter (MS) is a DX propagation mode which relies on reflecting signals off the brief ionised trail left by a falling meteor. These can last for up to a minute or more on rare occasions, but more usually for fractions of a second. This requires specialist operating techniques, relying nowadays mainly on the WSJT signal-processing software.

What are meteors? Meteors are particles of rocky and metallic matter ranging in mass from about 10^{-10} to larger than 10kg. About 10^{12} are swept up by the earth each day. At an altitude of about 120km they meet sufficient atmospheric resistance to cause significant heating. At 80km all but the largest are totally ionised, and this ionisation can be used to scatter radio signals in the range 10MHz to 1GHz.

A spectacular meteor trail.

Fig 4.9 shows the seasonal variation of meteor activity, based on a daily relative index.

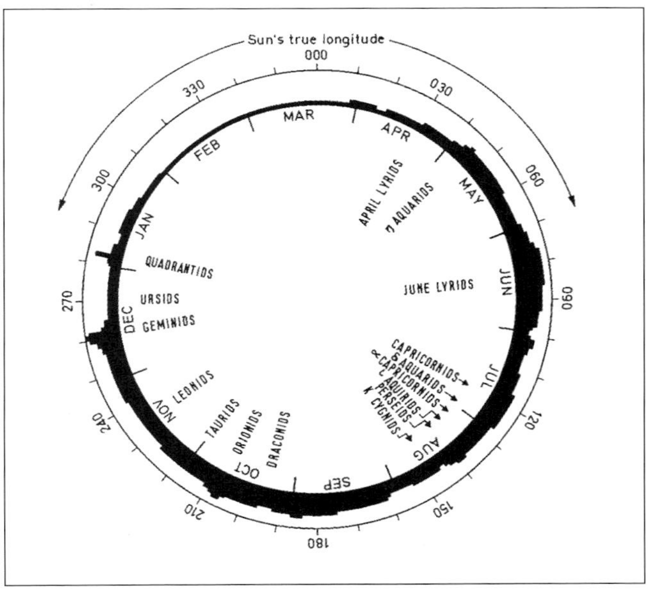

Fig 4.9: Seasonal variation of meteor activity.

Meteor scatter was once the province of specialist VHF operators but, with the advent of the WSJT software suite, it is now in daily use by many regular 6m and 4m operators with modest stations. Indeed, although meteor scatter propagation has traditionally been associated with the major meteor storms, the earth is constantly being bombarded by meteors or other cosmic debris and, using the WSJT software, it is rare that meteor scatter QSOs cannot be made. Signal strengths can be quite high and distances are comparable with those achievable via Sporadic E.

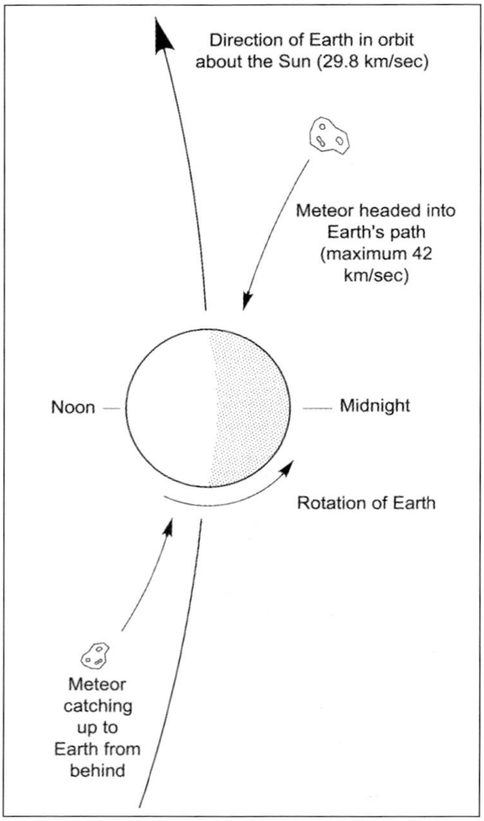

Fig 4.10: Relative velocity of meteors, as affected by the rotation of the earth (image courtesy of *ARRL Handbook*).

The problem lies in the fact that each meteor is very short-lived, so a contact can take a long time, passing just part of the overall exchange as each enhancement occurs. This is where the WSJT software makes life so much easier, as it handles the exchange of information and takes care of the precise timing that is required to effect a meteor scatter QSO. As such the operating technique is quite different to that normally used on the band and is described in detail in Chapter 6.

For the purposes of 6m and 4m operation, there are essentially two scenarios. One is to use regular meteor showers which are predictable and consistent. There are several websites which provide details of forthcoming meteor showers and, more importantly, give data on the timings and the radiant (i.e. the angle of entry into the atmosphere). By beaming at 90° to the radiant (in other words such that the meteors pass across the front of your antenna) you will maximise your chances of making a contact (for reasons which won't be discussed in detail here, the best beam heading is actually offset from this by 9° – 10°: the WSJT software tells you what offset to use). Other stations, to the far side of the meteor shower will be doing the same and, all being well, you will be able to work them by following the techniques described in Chapter 6.

Debris is constantly entering the earth's atmosphere and the second possibility is to use random meteors. The relative velocity of meteors and meteor-like material that meet the earth head-on is increased by the rotational velocity of the earth in its orbit around the sun. Fast meteors strike the morning side of the earth because their velocity adds to the earth's rotational velocity, while the relative velocity of meteors that 'catch up' from behind is reduced (see **Fig 4.10**). This makes dawn the best time of day to try random meteor scatter contacts, though this shouldn't preclude trying at other times of the day.

MOONBOUNCE (EME)

Not terrestrial propagation, but amateurs have, for many years now, been using the moon as a passive reflector of VHF signals. In theory, any frequency which is not reflected by the ionosphere but passes out into space can be used for moonbounce, also known as EME (Earth-Moon-Earth),

Beaming at the moon... The full moon, 2 September 2020.

communication. In practice, the lowest frequency in regular use is 6m, where moonbounce is possible only when Sporadic E and other enhanced propagation is absent. After all, if the ionosphere is reflecting 50MHz signals back to earth, they cannot be expected to pass through and reach the moon.

Path losses can be calculated with a reasonable degree of accuracy, allowing the station to be designed with sufficient antenna gain, transmit power and receive sensitivity to achieve two-way communications (depending on the station at the other end of the QSO too). Modern software-based systems, discussed in Chapter 6, have made EME accessible to many more amateurs.

Historically, EME required large antenna arrays and it was also necessary to point the antenna accurately at the moon and follow the moon as it moves across the sky. This is no longer the case because, as described in Chapter 6, the WSJT software brings 6m EME into the scope of more modestly-equipped stations. Nevertheless, the station requirements remain quite demanding.

Best results are achieved when the moon is just above the horizon, both because this does not require the antenna to be tilted and because your antenna will benefit from a few dB of 'ground gain' at these low angles.

Many factors apply in determining the best time for 6m EME contacts, not least because EME requires your signals to pass directly through the atmosphere in both directions whereas, as is clear from this chapter, there

EME DXpedition to the Seychelles by Lance, W7GJ, in September 2019 in which 108 EME QSOs were made.

are many ways in which those signals can be affected en route. EME is not going to work, for example, if a Sporadic E cloud reflects your signals back to earth.

At those times you are advised to enjoy the other propagation modes while they exist. But EME is a fallback when other propagation is lacking. At the time of writing, for instance, as mentioned elsewhere in this book, the tough ZL path has only been worked from the UK by EME, and a number of recent expeditions have made contacts via EME when other propagation modes have simply not been available.

TROPOSCATTER

Troposcatter propagation utilises the troposphere rather than the ionosphere and takes advantage of irregularities in the troposphere due to differences in the level of moisture, pressure, etc. Such irregularities are always present so troposcatter is available on a daily basis, but signal strengths are low and therefore a high-power station with a high-gain antenna system is required to make good use of this mode of propagation, which can be expected to support contacts up to distances of about 800km.

IONOSCATTER

While on the subject of scatter modes, for sake of completeness it is worth mentioning Ionoscatter. The ionosphere is far from being a homogeneous medium and while the ionosphere as a whole may not be dense enough to reflect 6m signals, occasionally those dishomogeneities may result in denser patches of ionisation sufficient to scatter signals back to earth. This can sometimes happen, for example, in the aftermath of Sporadic E.

It is worth saying that all scatter modes necessarily benefit from good antennas and plenty of power (commercial troposcatter links can use megawatts of ERP) because signals are scattered randomly in many directions so the losses over a given path will be high.

PROPAGATION INDICATORS AND BEACONS

One of the key elements of success on 6m and 4m is to be aware of likely propagation. This chapter has discussed the various types of propagation which occur and given a general idea of when those types of propagation might be expected. But a 'general idea' is not enough. You want to have a very specific indication of when a propagation opening is likely, as even the best forecasting tools can only give some idea of the probability of an opening.

Amateurs have installed beacon transmitters in many locations around the world and these are one of the most useful tools available. Nowadays it is trivial to program their frequencies into the memories of your radio and leave it to scan them on a rotating basis. Lists of 4m and 6m beacons in appear in Appendix E and Appendix F respectively.

As has been mentioned earlier in this book, an even earlier warning is available if you can spot signals from your target area on frequencies below 50MHz, especially if you know of several such transmissions on ever-increasing frequencies, so that you can track the steady rise of the MUF. It isn't possible here to give a list of transmissions that might provide this sort of indication as the list will vary from country to country.

In the past VHF TV transmissions were an excellent guide, especially if you had a TV on which to monitor them, as you could then see their ID and know where they were. Most of these have now closed down, so it is usually a case of finding other commercial transmissions in the 40 – 50MHz range in the area in which you are interested. There are plenty of frequency guides, both in paper form and on the web, which list such transmitters, and over time you will find out which ones work for you and gradually build up a list of them.

FURTHER SOURCES

There is a huge amount of material about VHF radio propagation, both professional and amateur, in books, magazine articles and on the Internet. It is almost impossible to know where to start, though some Internet browsing will give you plenty of pointers (for example, try searching for "50MHz propagation" using Google or any other search engine). Appendix B lists a number of useful sources.

One of the better amateur websites dedicated to VHF propagation is that of DF5AI **[6]**. But amateurs cannot compete with the professionals for real-time propagation data and you will want to bookmark sites like the excellent spaceweather.com **[7]** where you can find the latest solar data, auroral data and daily photographs of the sun. You can also sign up for space weather alerts to be e-mailed to you when significant solar events (such as a major flare or aurora) occur.

REFERENCES

[1] 'Sporadic E – A Mystery Solved?' Parts 1 and 2, *QST* October / November 1997.
[2] JE1BMJ SSSP: https://tinyurl.com/y48vnukk
[3] SolarHam: http://www.solarham.net/
[4] TEP bibliography: www.dxmaps.com/tepbiblio.html
[5] *Radio Auroras*, Charlie Newton, G2FKZ (SK), and Neil Carr, G0JHC, 2nd (updated) edition, RSGB, 2012.
[6] DF5AI: www.df5ai.net
[7] Space Weather: www.spaceweather.com

Chapter 5

Basic Operating

THIS CHAPTER TAKES A LOOK at the basics of operating under commonly-experienced propagation conditions. But it also covers how to find DX. Operating is much as it has always been (though the basics sometimes seem to be forgotten in the heat of a pile-up).

But what has changed out of all recognition over the last 20 years or so (mainly since the arrival of PacketCluster in the late '80s), and continues to do so, is the way in which DX stations are tracked down in order to get around to the serious business of actually putting them in the log. And the more recent but equally significant change is the explosive growth in digital modes, particularly FT8.

A WORD ABOUT MODES

If you are reading this book with a view to using 6m and 4m much in the way that you may have been using FM on 2m, that is to say simplex and via repeaters, then fine. There is something in this chapter about how to do so and you will enjoy plenty of activity on the mode. 4m in particular is home to many local FM nets using modified commercial equipment and is a great band for such activity.

On the other hand, if you want to chase DX, you will need to expand your horizons somewhat. Yes, when Sporadic E is in evidence you will be able to work out to several hundred miles on FM with loud signals and have a lot of fun in the process. But the serious DX will be on SSB and, under marginal conditions, CW (or, nowadays, WSJT, not just for meteor scatter and EME, but FT8 and FT4 (primarily) for weak-signal terrestrial working too). More about the latter in the next chapter. But the point to emphasise here is that, if you really want to catch all that DX, then ideally you will need to be able to work CW as well as SSB, although it has to be said that the arrival of FT8 has rather taken the impetus out of that. Nevertheless, CW operating is a useful skill to have. This can be quite a challenge at first, but is well worth the effort.

The good news is that CW decoders are improving all the time and many DXers make their first forays on to CW nowadays by using a decoder for receiving and a pre-programmed keyer (or maybe using the PC's keyboard) for transmitting. This set-up won't be effective for long 'ragchews' on CW but can be sufficient to copy a DXpedition station, to call him, see your callsign come back from him with your signal report, and to answer him with a pre-programmed signal report from you. That's all that's needed for a "good QSO". Though hopefully, over time, your own CW skills will improve and eventually you will be able to do away with the aids and send and receive as well as the best of them.

I should also mention the digital voice modes although they are not widely used on the 6m or 4m bands. There are three mutually incompatible modes at present, D-STAR, Yaesu Fusion and DMR. All have their merits and all allow not only direct two-way QSOs and repeater QSOs, but also access to international nodes via the Internet. That said, at the time of writing only one UK 6m repeater is internet-connected, a Fusion-capable repeater in Essex.

GETTING STARTED

The basics of operating on Six and Four are exactly as they would be on any other band and readers are advised to consult one of the standard texts such as the RSGB or ARRL operating handbooks. But it is worth reiterating some operating advice within these pages, with specific hints and tips as they relate to Six and Four. This chapter covers the basics that you will need to call upon in everyday operation, during Sporadic E and F2 openings, contests etc. The next chapter covers specialist operating, as it relates to EME and meteor scatter, where unique protocols have been devised to cover the nature of these modes and ensure that 'real' contacts take place (with meteor scatter, for example, as will be explained, a contact may take some considerable time due to the random nature of meteor trails. Fortunately nowadays, the QSO protocols are taken care of within the MSK144 software).

It also covers the use of FT8 and FT4, modes supposedly for weak-signal working although increasingly employed for day-to-day QSOs, even when propagation would support perfectly good CW or SSB QSOs!

The 6m band plans for the UK and the US are shown in Appendix E and Appendix F, while Appendix G is the RSGB 4m band plan.

On both bands it should be recognised that band plans vary from country to country, simply because national allocations vary, but the basic structure remains constant. In common with many other amateur bands, the bottom part of each band is reserved for CW operation, then SSB, then other modes including data modes and channelised FM.

FM OPERATION

Operation on the channelised parts of the bands is straightforward and will be familiar to anyone who has worked simplex or via repeaters on other VHF or UHF bands. Always listen before transmitting, keep your transmissions relatively short in case others in the area wish to make use of the channel, abide by your licence requirements for identification, and so on. These FM channels are primarily for day-to-day use, over essentially 'line of sight' distances (in the case of repeaters that means line of sight to the repeater). When the band is 'open', via Sporadic E or other long-distance propagation modes, it may be possible to work over greater distances on

FM. But at these times it is more appropriate to move down the band and switch to SSB or CW, modes which will cope better with the fluctuating propagation associated with long-distance working. In any case, don't try to access distant repeaters – repeaters are there to extend the operating range under flat conditions but if you try using them to work 'DX' you will simply deny access to those, such as mobile operators, who are trying to use them for their intended purpose.

Which takes us nicely on to operating in the lower reaches of the band. Obviously, some day-to-day working takes place on CW, SSB and especially FT8 (where considerable distances can be worked even under flat conditions), but it has to be said that these parts of the band really come to life during a lift in propagation or when there is a contest running. Contest operation will be covered a little later in this chapter and more fully in Chapter 9, while digital modes are covered thoroughly in Chapter 6, but first some thoughts on DX operating on SSB and CW.

DX OPERATING

The nature of 6m and 4m propagation is that openings can be very short, very focused and often characterised by rapid fading. Often you will first be aware that a DX station is being worked in your part of the world as a result of a 'spot' on the DX Cluster system or maybe by hearing a station within ground wave range of you working the DX. You will point your beam in the appropriate direction, tune to the DX frequency and start listening out for the DX station.

Before calling there are certain things you need to establish. These include determining whether the DX station is working 'split', i.e. listening on a frequency other than the one on which he is transmitting. If you are a regular HF band DXer this is a practice with which you will be very familiar. But if your background is largely on VHF, it may be new to you. The basics are covered in the sidebar 'Split Frequency Operation' on page 94.

Then listen to his rhythm and to the information he is exchanging. If you have little experience of working DX, it is easy to assume that you should make a long call: "Three Bravo Nine Charlie, Three Bravo Nine Charlie, this is Golf Three X-ray Tango Tango, Golf Three X-ray Tango Tango, over". But this is a bad idea. With rapid fading you may only have a matter of seconds to make a complete contact before his signal fades into oblivion. On the other hand, if he is consistently loud, he will probably have a big pile-up of callers and want to get through them as fast as possible. He already knows his own callsign, so no need to give that, and give yours just once, or twice at the most, before listening to see who he has responded to. Hopefully, it will be you, in which case you can respond with his report, "3B9C thanks, 59".

Many VHF operators like to give their QTH locator, but only do this if the DX station appears to be giving this information and asking for it from those he is working. Many DX stations will consider it a waste of time, simply

SPLIT FREQUENCY OPERATION

(While much operation currently takes place on FT8, split-frequency operation can still be expected on CW and SSB when conditions permit and a DXpedition operation decides to use those modes)

Nowadays, with the huge interest in DX chasing, most DXpeditions and many DX operators choose to operate split frequency. The concept is simple. By transmitting on one frequency and listening on another, callers can hear the DX station clearly, rather than through a mass of other callers, and therefore know when to go ahead with their contact and, more importantly perhaps, when to remain silent while another contact is taking place. And rather than listening on a single frequency, the DXpedition may choose to listen over a range of frequencies, thereby making it easier to pick out callers.

If you hear a DX station or expedition making plenty of contacts, but you can't hear the callers, then the chances are that you are listening to a split-frequency operation. Indeed, before calling any DX station on his own frequency, it's always worth waiting a moment to determine whether he is working split. Otherwise, if you call him co-channel, you risk the wrath of others who have been waiting and will almost certainly inform you of the error of your ways!

If a DX station is working split, what do you do? In years gone by many amateurs may have had a problem in that their transceiver was only capable of transmitting and receiving on the same frequency. In practice most transceivers came equipped with RIT and XIT (Receiver and Transmitter Incremental Tuning), allowing for a limited degree of split operation (typically up to 10kHz between transmit and receive frequencies). Most modern transceivers go one step further and have two quite separate VFOs, giving total flexibility (though it's worth practising split operation in the peace and quiet of your own shack before you try it on the air and end up pressing the wrong buttons!) Most radios also allow you to check your transmit frequency, which is handy, because as well as listening to the DX station, half the trick can be finding the people he is working, and putting your transmitter on that frequency. The more expensive radios go one step further still, with a second receiver. With that capability, you can listen to the DX station in one ear of your headphones (I never use a loudspeaker for anything other than casual ragchews, and I suspect most DXers are the same) and the pile-up in the other ear. Then you know exactly what is going on.

Always listen to the DX station carefully. Often he will announce his receive frequency, and may also be giving other instructions (such as "UK stations only") which should always be followed carefully to avoid creating unnecessary interference and slowing things down. If no listening frequency is announced, the general rule is to call about 1kHz up on CW, or 5kHz up on SSB. But, again, a few moments' listening should, in any case, quickly allow you to find the callers he is working. Listen for a little longer and you may also determine a pattern. For example, does the DX station always respond to callers on the same frequency or does he, for example, listen a little higher up the band after each contact, finally dropping back down the band and starting the whole process anew? Does he respond to stations only giving part of their callsign or does he, like many DX operators, only respond to callers when he has their full callsign?

And so on. Some intelligent listening can pay dividends, compared with simply calling at random. This is how experienced operators running low power can often get through more quickly than less experienced operators running high power (No surprise there, I suppose. Experience counts for more than brute force in most competitive sports and activities).

Unless the DX station is specifically taking 'tail enders', i.e. stations who call as the previous QSO is coming to an end, never call over the top of a contact in progress but wait until the DX station signs and calls "QRZ?" or similar. Otherwise chaos ensues.

reducing the number of contacts they can make during what may be a brief band opening. This information can easily be obtained from the web or from an exchange of QSL cards. And only repeat your own callsign if he has it wrong and you need to make a correction.

On WSJT, the QSO protocol is taken care of by the software, so you just have to live with that and trust that the propagation remains stable enough for the QSO to be completed. So, much of the following applies to SSB and CW operating – WSJT setup and procedures are dealt with in detail in the next chapter.

OPERATING DURING A SPORADIC E OPENING

Sporadic E openings on Six and Four are relatively common each summer and at some other times of the year, and propagation can exist for several hours at a time. Using Sporadic E to make QSOs is therefore no different to how you would conduct a QSO via any other day-to-day mode of communication.

Avoid using the so-called calling channels – there is no hope of establishing an Es contact and then changing frequency as is normal practice with other propagation. All too often a station fades out during a contact and there is little you can do other than move on or, if it's a rare one you need, wait around in the hope that propagation will return. The usual rules for a contact apply – it is only valid when a confirmed exchange of calls and reports has taken place – if you haven't heard him confirm your report than there is no way you can be sure you are in his log.

Hopefully, the propagation will reappear, or maybe you'll just have to wait until another day. The cloud of E layer ionisation, which may be relatively small in area, can move very rapidly and also tilt. For example, from the UK, Maltese stations may be very strong one minute then fade out to be replaced by Sicilians which are in turn replaced by Sardinians, all in the space of a few minutes.

CQ calls should be brief, perhaps a maximum of 10 seconds, and used with discretion; if an operator is in an area of high activity, any DX station is likely to have many UK stations calling him and be unlikely to reply to a call. Conversely, if an operator is isolated, he may have an opening all to himself and a pile-up of DX may call him. In this case he will want to work as many as possible in the time available, so he will probably just exchange reports with each station maybe giving the locator on every third or fourth contact (the waiting pile-up will have heard it already).

However, always remember that Sporadic E can be very localised and it is possible for a DX station who is S9 to one operator to be inaudible 10km away – and that a CQ call may well pay dividends, especially just as the band is opening. This variability, as discussed in the propagation chapter, can be even more pronounced on multiple-hop Es: even more reason to keep contacts short and sweet. When multi-hop Sporadic E is in evidence (and, for example, it often occurs late in the day European time in the

Z6 (Kosovo) was added to the DXCC list in 2018 and makes for a nice catch on 6m via Sporadic E.

summer, to the Caribbean and North America) signals will be far weaker than with single-hop, so all the foregoing advice is even more relevant. Short, snappy contacts, not only to ensure that your own contact is good before conditions change, but also to allow as many others as possible to make the contact during what might prove to be a short and one-off opening.

AURORAL OPERATING

The beacon stations closest to the auroral zones are usually the first auroral signals to be heard in the south. The Faeroes beacon OY6BEC (50.4022MHz and 70.0354MHz), the Swedish beacon SK3SIX (50.468MHz), the Danish beacon OZ7IGY (50.471MHz and 70.021MHz) or the Finnish beacon OH1SIX (50.012MHz) are all possible indicators of auroral activity in Europe, along with other similar beacons in JX, JW, LA, OX, OZ, TF, etc. Similarly, the various northerly 4m beacons that now exist.

Auroral events are usually first noticed by amateurs in Norway, Sweden, Finland and the Baltic states, who will be on the band making contacts before the auroral reflections extend to the south. Due to their northerly locations, they see far more visual auroras and participate in more radio events than stations in, say, southern England. It is rare that auroral propagation will extend into southern Europe, but stations in that part of the world benefit more from other propagation modes. This said, amateurs have noticed that the larger the change in geomagnetic activity, the farther south the area of auroral ionisation extends.

If you hear someone calling "CQ aurora" and you have not operated in an auroral opening before, resist the temptation to call indiscriminately, and listen only. There are going to be hundreds of future auroras and this is your chance to learn the entirely new operating techniques required for auroral contacts. First select horizontal polarisation and beam between north and east at 45° (later the centre of propagation may move westwards). Tune the beacon band; it will take some time to get used to the rough sounding keying of the beacons which will be slightly off their usual frequencies due to Doppler shift. Check which beacons are audible. If, for example, you hear one of the German beacons, the aurora is extending at least as far as Germany, and is therefore a large-scale event which will probably last for a few hours and may repeat later in the evening. Turn the beam between north and east on each beacon heard and it will be noticed that different beam headings give peak signals for different countries. Generally, the farthest DX is worked with the beam well off a northerly heading.

Next tune the SSB section of the band and try listening to an experienced local station who is working auroral DX. Due to the distortion he will be speaking slowly, using correct phonetics and possibly end-of-transmission tones. Remember that you will hear the local station direct but the DX stations will be replying via the aurora and will be slightly off his frequency.

A typical SSB auroral contact starts like this:

"CQ aurora, CQ Aurora, GM8FFX, Golf Mike Eight Foxtrot Foxtrot X-ray calling CQ aurora . . ." (repeated slowly several times) "and GM8FFX listening." pip (end of transmission tone).

"GM8FFX, GM8FFX, LA2PT calling. Lima Alpha Two Papa Tango, LA2PT calling GM8FFX . . ."

End-of-transmission tones which give a low-frequency pip are very helpful in auroral openings when signals are weak – the tone readily identifies the end of each station's transmissions and could be a 'K' tone or 'pip' tone. Incremental receiver tuning (IRT or RIT) is a 'must' for auroral reception as the amount of Doppler shift often changes in the middle of a contact.

It will soon be seen that SSB contacts are difficult due to the distortion, and contacts tend to be limited to exchanging reports and maybe locators. Many amateurs also exchange and log the beam headings used at both ends of the contact as a study of these figures can reveal the particular area of ionised E layer being used.

During weak auroral events SSB operators in England, Northern Ireland, North Wales and Scotland can work each other and operators with better facilities can contact Norway and Sweden. During strong auroral events SSB stations all over the UK and north-west Europe can work each other.

A new operator listening in the CW section of the band will hear a great many rough-sounding hissing CW signals during a strong event. Experienced operators are used to the strange AC-sounding notes and contacts are completed quickly and efficiently. The letter 'A' is added when calling CQ and is also added after the readability and signal strength report, in place of the normal tone reports which are not sent during aurora openings as no signal sounds T9. The best auroral DX is always worked on CW, just as on other propagation modes; CW is easier to copy in weak signal conditions and contacts are therefore completed much faster.

Auroral openings can occur in three separate phases in a single 24-hour period. The first phase can start as early as 1300 local time but usually takes place between 1500 and 1900 local. The second phase can occur between 2100 and 2300 local and a third phase can run from after midnight till 0600. Very few auroral contacts take place around 2000 local and there is often a fade-out between the evening phase and the after-midnight session. Some auroras have no afternoon phase and start in the evening, often continuing again after midnight. Some auroras have no afternoon or evening phase, only starting after midnight. These are almost always weak events, sometimes heralding a larger occurrence on the next day.

Due to the Doppler shift and distortion on signals, auroral contacts can only be made on CW and SSB. While some stations have attempted to use FT8 mode during auroral openings, this really doesn't work due to the auroral flutter and distortion.

High power is not essential but helps greatly in weak events. During strong events almost anyone can participate. Signals reflected from the auroral curtain do not change polarity, and high-gain horizontal antennas give the best results. Operators who are blocked to the south and south-east enjoy auroral openings as they can work stations that are normally unheard due to the obstructions, by beaming well to the north of the direct path.

The amount of Doppler shift is proportional to the frequency band in use. For this reason, auroral signals on the 50 and 70MHz bands are easier to read and have less distortion than on the 144MHz band. VHF operators coming to 6m from 2m will already be familiar with auroral propagation. HF operators may have experienced such propagation on 10 or 12m. But there is no mistaking auroral propagation when you run across it, the sound of signals being so different to any other propagation mode.

OTHER MODES

The WSJT software suite for weak-signal working on 6m is covered in the next chapter, as it is a major topic in its own right. There is nothing to prevent you using other modes on 6m, for example RTTY (teletype) and SSTV (slow scan television) when conditions are suitable, and some operators do exactly that. The use of these modes is well covered in other literature such as the *RSGB Operating Manual*. To set up for them, you will need your PC linked to your radio and suitable software such as MMTTY for RTTY and MMSSTV for SSTV. Remember, though, to observe the band plan, which has designated calling channels for these alternative modes.

CONTEST OPERATING

A short word about contest operating, which is covered in a little more detail in Chapter 9. You may think of yourself as a DXer, not a contester. But the serious DXer will want to participate in 6m and 4m contests, perhaps not to achieve a high score or maybe not even to send in an entry at all (though it will always be welcomed by the organisers as a help to checking other logs). The reason to participate is simple. Contests bring activity on to bands which otherwise might be silent. And that activity can do two things for you. Firstly, it may highlight propagation paths that are open and of which you are only aware because of the activity taking place. And, secondly, it may bring on to the band activity from countries, squares, islands, cantons or other 'counters' that you are chasing, especially as contests are often an excuse for some entrants to go out and set up portable stations for the occasion. So, do be aware of contest activity and be ready to take part even if you don't consider yourself a contester *per se*.

FINDING THE DX

A lot has already been said about the types of propagation you will find on Six and Four and how to spot possible propagation opportunities. But there is much more to working DX than simply being around when suitable propagation occurs. The secret is to have access to good information. For example, the various amateur radio magazines and bulletins have always made a point of carrying news about forthcoming operations, as well as wrap-ups of what has been on and worked. Nowadays the problem, if anything, is one of information overload. Your national magazine, such as *RadCom* in the UK and *QST* in the USA, will have a VHF column, carrying useful information, but their lead times mean that breaking news will often arrive too late to be carried.

To supplement these sources, DX chasers turn to Internet-based bulletins such as *The Daily DX* **[1]**, a subscription service by W3UR, the *OPDX*

The Daily DX is one of several Internet-based DX bulletins.

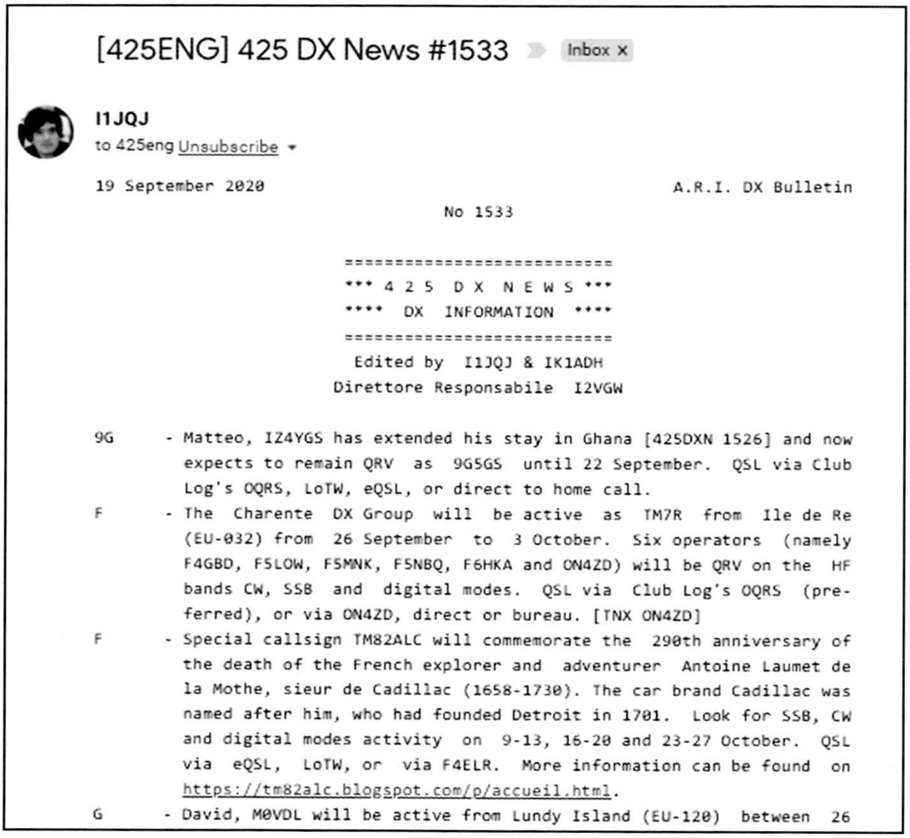

[425ENG] 425 DX News #1533 Inbox ×

I1JQJ
to 425eng Unsubscribe

19 September 2020 A.R.I. DX Bulletin
No 1533

===========================
*** 4 2 5 D X N E W S ***
**** DX INFORMATION ****
===========================
Edited by I1JQJ & IK1ADH
Direttore Responsabile I2VGW

9G - Matteo, IZ4YGS has extended his stay in Ghana [425DXN 1526] and now expects to remain QRV as 9G5GS until 22 September. QSL via Club Log's OQRS, LoTW, eQSL, or direct to home call.

F - The Charente DX Group will be active as TM7R from Ile de Re (EU-032) from 26 September to 3 October. Six operators (namely F4GBD, F5LOW, F5MNK, F5NBQ, F6HKA and ON4ZD) will be QRV on the HF bands CW, SSB and digital modes. QSL via Club Log's OQRS (preferred), or via ON4ZD, direct or bureau. [TNX ON4ZD]

F - Special callsign TM82ALC will commemorate the 290th anniversary of the death of the French explorer and adventurer Antoine Laumet de la Mothe, sieur de Cadillac (1658-1730). The car brand Cadillac was named after him, who had founded Detroit in 1701. Look for SSB, CW and digital modes activity on 9-13, 16-20 and 23-27 October. QSL via eQSL, LoTW, or via F4ELR. More information can be found on https://tm82alc.blogspot.com/p/accueil.html.

G - David, M0VDL will be active from Lundy Island (EU-120) between 26

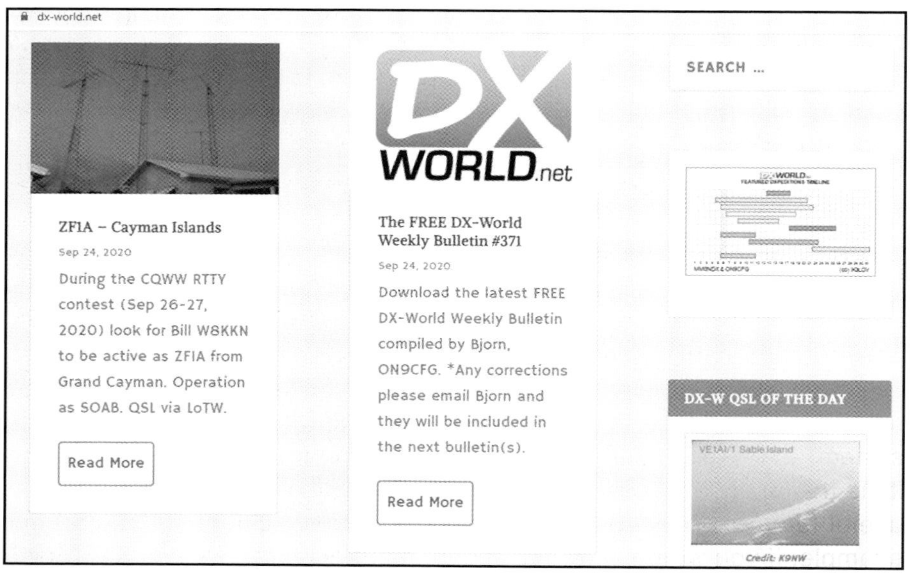

Top: The 425 DX News is sent weekly as an e-mail. Below: The DX-World.net website is a good source of DX information and is updated regularly.

Bulletin **[2]** and *425 DX News* **[3]** (these latter two are free, but weekly rather than daily). And the DX-World website is an excellent source of such information and regularly updated **[4]**. All of these sources carry 6m information.

Most expeditions will also put up a website beforehand, with information about frequencies, times of operation, callsign etc. This may well include contact details so that you can contact the DXpedition either directly or via a pilot station, for example to arrange skeds or pass details about such issues as frequencies to be avoided, etc.

But the real revolution has been in real-time information sources. The way you actually make a contact is much as it has always been. The way you find the DX station has changed out of all recognition unless you are one of the (increasingly few) who refuse to turn to external sources and simply enjoy listening to the band (white noise for much of the time!) until something shows up.

Let's look at a few scenarios to show how a serious 6m DXer might ensure that he doesn't miss the current DXpedition (and the same might apply to 4m, as more DX and DXpedition stations start to appear). The first trick is to track the MUF as it starts to rise, whether by F2, Sporadic E or other means. For this, increasingly, serious DXers are turning to wideband SDR receivers able to monitor several frequency segments simultaneously. They will be able to watch, on the waterfall display, as signals start to appear at ever increasing frequencies. Some will ensure that their wideband receiver is supplemented with a wideband antenna (probably a log periodic) as their 6m antenna will be too narrow in its bandwidth to be an effective antenna across, say, the 30 – 80MHz spectrum.

Then, of course, there are the various Internet-based alerting systems that will let you know when the DX is being heard and worked, albeit not necessarily at your location.

INTERNET-BASED OPERATING AIDS

The most popular source of real-time DX information has to be the Cluster system (formerly PacketCluster, but now accessed by most people via the Internet). You are probably already familiar with this global network, whereby amateurs around the world make 'spots' of what they are hearing, this data then being available to everyone connected to the network. The Cluster network also carries solar data (flux, A and K indices), which can be a useful indicator of possible propagation. The Cluster network can be accessed using Telnet (a communications protocol built into your PC's operating system, and also integrated into many logging programs) or you can access it via the DX Summit website **[5]**. DX Summit also has a number of other useful features, such as being able to search for 'old' spots so that, for example, if you were out and missed a band opening, you can look back and see what was heard and by whom, which may give you some clues as to best times and frequencies to find the specific station(s) you are chasing.

Spotter	Freq.	DX	Time	Info	Country
EA4LL-@	50305.0	EA4LL	19:41 23 Sep	USB	Spain
IW8EHC	50313.0	IW8FEN	19:13 23 Sep	FT8 -07dB from JN70 1324Hz	Italy
HF5ZM-@	71333.0	IY8EJ	16:49 23 Sep	grazie	Italy
GM4UYE	50313.0	F8ZW	14:56 23 Sep	Tnx QSO Jean-Paul 73	France
SM2CEW	50417.0	OH9SIX	14:19 23 Sep	57a qtf 015	Finland
G4FKA	50313.0	DJ1AA	13:40 23 Sep	CQ	Fed. Rep. of Germany
KS0AGY	71436.0	IY6SB	12:33 23 Sep	Award Marconi Day	Italy
G4FKA	50313.0	EI2CN	12:29 23 Sep	CQ	Ireland
KD9VV	50260.0	NF4A	12:23 23 Sep	EN71<MS>EM70	United States
NL8992-@	70095.0	PA75AG	12:05 23 Sep	PI4 12 dB Q=100	Netherlands
F5VCR-@	71989.0	F4FET/P	11:58 23 Sep	AT134 ILE AUX RATS	France
W4HLR	50145.0	K4TR	11:54 23 Sep	EM56JB<MS>EL88TN 5/3 tnx	United States
K0KKO	50145.0	W9DR	11:49 23 Sep	57 into EM8	United States
K0KKO	50145.0	W4ICU	11:47 23 Sep	53 into EM8	United States
KD9VV	50260.0	KA1W	11:45 23 Sep	EN71<MS>FN31	United States
KD9VV	50260.0	VE1JF	11:42 23 Sep	EN71<MS>FN74	Canada
W4HLR	50145.0	W4ICU	11:41 23 Sep	EM56JB<MS>EM70KP tnx	United States
K0KKO	50145.0	K4WMS	11:39 23 Sep	55 into EM48	United States
WB8AKW	70026.8	3D2AG	11:38 23 Sep	Up 1	Fiji

The DX Summit Cluster, filtered to show 6m and 4m spots only on 23 September 2020.

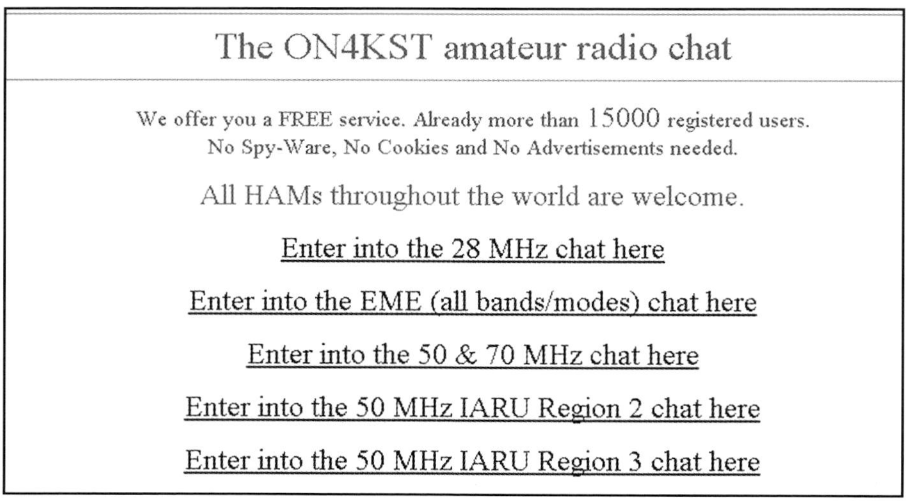

The ON4KST chat rooms cater for specialist DXers and now has more than 15,000 registered users.

Although you can set DX Summit to show only 6m or 4m 'spots', you may prefer to go instead to a site specifically dedicated to 6m, several of which exist around the world. Equally, you may choose to use the VE7CC Telnet software and select to see only 6m or 4m spots.

The other invaluable facility is the ON4KST chat room **[6]** (and a few other, similar, chat rooms elsewhere in the world). The ON4KST site has several 'rooms' including one for topband (160m). But the one that will interest readers of this book is that for 6m and 4m. Here you will find a group of like-minded enthusiasts gathered to exchange information, set up JT65A or MSK144 skeds, and generally chat about their favourite band. But the site also shows real-time propagation maps (continually updated from DX Cluster spots) and much more, making it a 'must have' tool for 6m stalwarts.

The danger with a site like ON4KST is that there is always a temptation to go beyond simply setting up a sked to the point where you start asking your potential QSO partner such things as "Did you get my report?" or even "Did you get the 57 report I sent you?" which makes any resulting 'QSO' a nonsense. Like any tool, used sensibly it is a valuable aid, but take care not to abuse it.

But this is by no means where the story ends. Nowadays there is a range of tools that draw data from the Cluster or Telnet nodes in a way which is much more useful to you as the needy DXer. For example, most logging programs will take spots from the Cluster network and filter them by DX callsign, country of origin and other algorithms which you, as the user, can define. This means that you only become aware of 'spots' which are of interest to you. Let's say you know (to take the 2012 example) that D64K is active and has a serious 6m enthusiast as part of the expedition team. You might set the filters on your software so that you only see spots for D64K, and only if the station making the spot is in your own CQ zone.

But during the course of a two-week expedition there may be just one or two fleeting openings to such a station (in the case of D64K from the UK that was indeed true although, fortunately, one of them was on a Saturday when many 6m DXers were at home). Can you spare two weeks sat in front of the radio 'just in case'? Probably not. In the past a network of friends might have agreed to alert each other by telephone in such an instance. Later, as technology advanced, some took to carrying pagers to be alerted to DX activity.

Just to mention a few other facilities available to you, although the range of tools is increasing almost by the day. Firstly, CW Skimmer software by Alex Shovkoplyas, VE3NEA **[7]**, will decode CQ calls on CW and present them on screen. If you run the software at home you can monitor the whole of the 6m band (or maybe even 6m and 4m) at the same time, seeing who is active on the band and audible at your location. Many current transceivers have a wideband output before the roofing filter to allow this, as well as to drive external devices such as panoramic adaptors. SDR transceivers such as the FlexRadio range cover 6m directly (those in the Flex 6000 range also

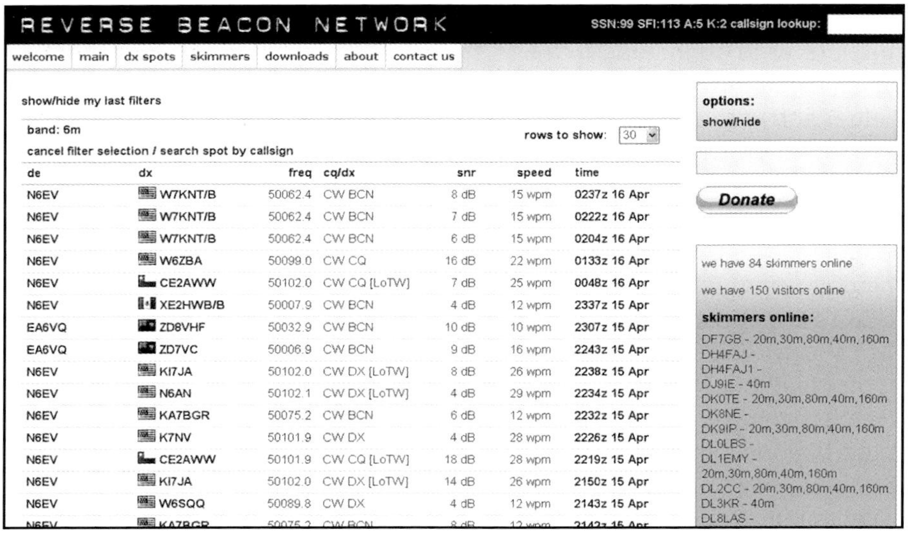

The Reverse Beacon Network allows users to monitor the outputs of 'CW Skimmers', either locally or wherever the user so chooses.

include 4m coverage). Again, this has already been discussed in Chapter 2 in the context of equipment selection. But, whether you have a local Skimmer or not, the international Reverse Beacon Network (RBN) **[8]** allows you to monitor the outputs of Skimmers elsewhere in your own country or farther afield (your choice). You can either sit watching the RBN output, or have it filtered by your logging or other software before sending you an alert to your VHF handheld, Smartphone or other portable device (or even operate an audible alarm around the house should you so choose!)

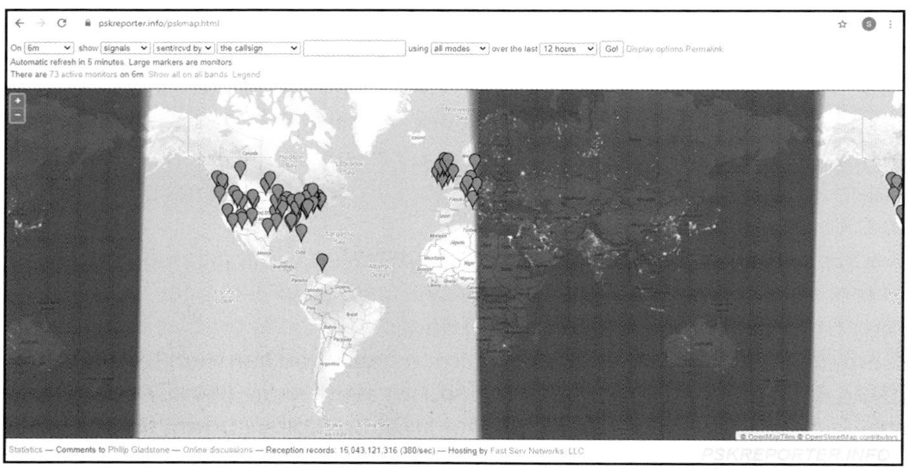

PSKreporter screen shot, showing all 6m spots at 1700UTC on 24 September 2020. The screen is updated every five minutes.

Perhaps most useful to the WSJT (FT8 and FT4) enthusiasts is PSKreporter **[9]**, which shows spots for all bands and modes so, for example, you can choose only to see FT8 spots on 6m. And you can use it in various ways. For example, you can use it to see where you are being heard (and, as Frank, W3LPL, recommended in a recent presentation, try various audio tones in the FT8 bandwidth to see which is getting through best at you desired target area). Or you can use it to see what is being heard close to your QTH (assuming there is a receiving site close to you). JTAlert is another useful program that works with WSJT to track awards, alert you to needed contacts and more. It is covered in more detail in the next chapter.

Having access to all these tools may seem like a form of cheating and there is nothing that says you have to use them. But there are many advantages. Look at it, for example, from the point of view of the DXpedition itself. Imagine that someone has spent quite a lot of money to haul equipment to the Caribbean for the summer Es season. The good news is that if and when an opening to, say, Western Europe occurs, then there will actually be people on the band to work who might otherwise have been blissfully unaware and carried on gardening, shopping or whatever else was on their agenda for the day. And such tools have also helped to understand propagation better, as an automated station can, for example, monitor beacons across Africa and elsewhere 24 hours a day, when the chances of someone listening on the right frequency at the right time otherwise are probably very low.

While on the subject of the Internet, in the UK, while there is no 4m club as such, there is an Internet reflector for 4m enthusiasts **[10]**. This is perhaps a good point at which to mention that there are nowadays, as you might expect, a number of Facebook groups for both 50MHz and 70MHz enthusiasts, some Open and some Closed. It's certainly worth investigating these if you are a Facebook user.

The range of Internet-based tools will undoubtedly continue to proliferate. There are, for example, mapping sites that overlay DX spots on to world maps to show recent 6m propagation, which gives you an almost instant view of, for example, the location of Es clouds. Nowadays there are facilities such as remote receivers through which you can listen to the band from various locations around the world. This can be fascinating in terms of hearing what 6m sounds like from elsewhere but, again, should not be abused (for example, by listening to a DX station that you cannot hear at your own station but raising him from your home simply because you have more transmit power available than he has).

It's a matter of ethics, something which has always been relevant to amateur radio, where we conduct our operations from the privacy of our own homes in a way which cannot easily be overseen by third parties (such as those from whom we might want to claim awards). But the Internet has dramatically increased the possibilities for cheating, so the ethics question is more pertinent than it has ever been.

VOLUNTARY OPERATING CODE OF PRACTICE FOR SIX METRE OPERATORS

as issued by the UKSMG in conjunction with JAROC, HARDXA, JAROC, SixItaly, DRAA, LABRE-SP and SSA, and reproduced with kind permission of UKSMG (Note that this has not been updated since 2012)

6m AS A DX BAND: 6m is a DX band just like any other of the amateur radio high frequency DX bands and it, along with other 6m operators, should be treated with respect and tolerance. Note: Please remember in Europe that French operators are not allowed below 50.200 so local QSOs held just above 50.200 could affect their ability to work DX.

LOCAL BAND PLAN: Always respect your local band plan. In Europe this is issued by the IARU. **LOCAL QSOs:** Do not cause nuisance and disturbance to other dedicated 6m local and overseas DX operators with local QSOs within the 50.100MHz to 50.130MHz INTERCONTINENTAL SECTION. If you do wish to local ragchew, it is recommended that you do this above 50.250MHz where interference will be minimised.

LEARN TO LISTEN: Most 6m DXers spend less than 5% of their time transmitting while 95% or more is spent listening and observing changing band conditions and propagation modes. Learn to recognise propagation mode characteristics and when the band is likely to be showing signs of an opening. This will be far more effective than just calling CQ DX at random and *ad infinitum*. **50.100 - 50.130 INTERCONTINENTAL SECTION:** The INTERCONTINENTAL SECTION is a widely accepted concept and should, in principle, be used for Intercontinental DX QSOs only, especially the 50.110 calling frequency as discussed below. The definition of what constitutes a 'DX' station naturally lies with an individual operator, especially when a particular station within your own Continent constitutes a new country! The 50.100 - 50.130 INTERCONTINENTAL SECTION should only be used for QSOs between stations in different continents or where the station is outside the range of single- hop Es propagation i.e. roughly 2400km or 1500 miles. We would ask you to think carefully before having any intra-European short distance QSOs in the INTERCONTINENTAL SECTION. For those of us in Europe, this is especially important in periods of multiple-hop Es or F2 propagation to avoid burying Inter-Continental DX QSO opportunities under a layer of European QRM.

PLEASE BE SENSIBLE and avoid local QSOs in the INTERCONTINENTAL SECTION if at all possible! As the INTERCONTINENTAL SECTION is heavily used, always listen before you call and always ask if the frequency is being used before you transmit (this should be done on any frequency anyway). Just because YOU can't hear anything, it does not mean that the frequency is not occupied or some rare DX is using it. Remember that operating etiquette calls for you to ask if the frequency is occupied BEFORE calling CQ.

50.110 INTERCONTINENTAL DX CENTRE OF ACTIVITY: The Intercontinental DX Centre of Activity is 50.110MHz. This should be used for long range DX contacts and such contacts should normally be intercontinental in nature. If a local station returns to your CQ, move quickly to an unused frequency above 50.130MHz. Do not use the DX calling channel for testing or for tuning up your radio / antenna. Do not encourage pile-ups on '110. If you have a successful CQ ensure that you QSY elsewhere in the band.

50.110 CQing: LISTENING is the first rule of working rare DX on 6m. So think twice before calling CQ on 110. It would be stupid to say that you shouldn't call CQ but please remember that this is a shared frequency so your reputation will be on line if you insist on calling CQ unceasingly every minute of the day or throughout an opening - even if you do say "CQ DX only" or "CQ outside of [my continent] only". The occasional CQ is good as it can discover an unrecognised opening.

If you are a 6m DXer and have been intensely listening for weak exotica for hours on 110 and up pops a CQ caller, rather than ask him rudely to clear off, ask them POLITELY to QSY and TELL THEM WHY OR WHAT YOU ARE HEARING OR LISTENING FOR, and please GIVE YOUR CALLSIGN .

Of course, this applies equally well to any frequency on 6m. Most operators are sensible and will do so, probably because they would like to work the DX themselves! Conversely, if you call CQ or are occupying 110 and someone asks you politely to QSY and GIVES YOU A REASON, do so without arguing about the rights of doing so: remember that that you share this resource with thousands of other operators. If you really must call CQ on 110, think twice, listen for five minutes, cross your legs, count to 100, and if the overwhelming desire is still there go ahead and CALL - but keep it short! At the end of the day the choice is yours and yours alone. Don't forget to QSY when successful unless it is inter-continental DX!

QSO TECHNIQUES: Many operators do not take the

time to learn how to DX, develop QSO skills and techniques and jump right in. This is not to be recommended as typical 6m propagation does not allow wasting of time during DX QSOs due to the nature of propagation of the band (borderline HF / VHF). Openings could be very short in time duration and DX stations wants to work as many callers as they can during an opening.

Basically, follow the style and take the lead of the DX operator in providing information. Otherwise keep it simple and to the point as there are other stations who are also waiting in line for a QSO with the DX station. Do not waste time in exchanging unnecessary information such as locator codes, names, QTH, equipment, weather and so on. Just exchange your callsigns and confirm your signal reports and move on to allow other DXers to have their QSOs. Leave out all the extra information (such as Maidenhead squares) unless it is requested. Many opportunities to work a DX station are extremely short and if your operating practices prevent others from working the station it will be remembered by those who missed out for a long time. Next time it may be you who misses out.

FREQUENCY CONFLICTS: With the quickly shifting propagation as regularly encountered on 6m, it is quite possible that two stations who have been occupying a frequency for several hours running pile-ups without hearing each other, to suddenly find themselves in a clash. In these circumstances, operators should mutually resolve the situation as quickly as possible to avoid conflict. It should always be remembered that no individual operator 'owns a frequency', even if you have occupied a frequency for several hours.

DX PILE-UP OPERATING: Working and breaking DX pile-ups can be as frustrating an experience on 6m as it is on HF. Manners and good operating are very important. You should listen to the DX stations carefully and not continue to call if they request a particular country or prefix to go back to them if that is not you. You should always go back with your complete callsign, give it quickly and give it only once. There is nothing more frustrating and aggravating for others in a pile-up for you to double with the DX station and miss who they going back to. Of course, you should NOT call if you cannot hear the DX station! If a QSO is uncompleted due to QSB or QRM, don't continue to try and complete the QSO to an excessive degree, use your judgment and call back later. It is likely that others are hearing them OK and can complete a QSO. Take the lead from the DX station and don't call back immediately if they are working someone else. The message is simple, try to avoid calling over the top of the DX station - it does you no good and just upsets your fellow DXers.

SPLIT FREQUENCY OPERATION: When a DX station creates a large pile-up of stations all calling him on their own operating frequency (simplex operating) it creates tremendous QRM problems for those calling and the DX station. Under these circumstances, it is recommended that the DX station uses split operating; that is transmitting on one frequency but listening over a range of frequencies above the frequency being used by the DX station. This mode of operating will significantly increase the QSO rate of the DX station. However, split operating on 6m can cause TREMENDOUS interference with other DX operators who, through no fault of their own, are running a simplex pile-up in the same split-frequency section of the band. To minimise this interference, it is recommend that a maximum split of 10kHz (definitely NOT 100kHz) is used.

DUPLICATE QSOs: It is always tempting to call a rare DX station every time you hear it. This should be avoided as it means that you taking away the opportunity for the DX station to work a new station and give them their first QSO with the DX country. Use your judgment if the DX station is known to be rare! Conversely, a quick call can sometimes be useful if no one else is going back to the DX station to show that there is propagation.

CW OPERATION: CW is probably the major mode of operation on 6m due to the usually weak nature of many real DX openings. Do not call CW DX stations using SSB as they will not be able to hear you and you will be causing severe interference to other CW DXers trying to work the station. The contrary is true as well, if you cannot break a SSB pile-up using SSB then do not call using CW!

FM QSOs: All FM transmissions should be made above 50.500MHz for the obvious reason that FM is wide band and could wipe out weak DX signals. There is no acceptable reason to transmit FM below 50.500MHz, as there is plenty of spectrum allocated for this purpose.

Note: This Code of Practice may be amended by UKSMG, JAROC and HARDXA from time to time to reflect current licensing conditions and operating practices.

ETHICS AND OPERATION ON SIX AND FOUR

Firstly, there are those things which are illegal, either because they contravene our licences or because they contravene the rules of specific awards or contests that we may be interested in pursuing. Then there are those things which are legal but, for various reasons, may be acceptable to some operators but not to others.

Things Illegal

Clearly some actions would be quite illegal and to increase your standing in Six metre circles by such conduct would do you no favours at all and, in some cases, could lose you your licence. Examples might include operating with excess power, operating outside your allocated frequencies, or asking someone else to operate your station under your callsign in your absence (allowable in some countries but not in others). Another instance would be using a remote station in another DXCC entity, closer to the DX, but signing your home callsign. Do any of these things and it may increase your score, but would you gain any satisfaction from so doing?

Then there are actions which, while of no significance to the licensing bodies, are considered out of bounds within the amateur radio community. One example would be submitting forged QSL cards for awards. This is easily done – one friend of mine has quite easily obtained blank QSL cards from over 170 DXCC countries (samples from printers, cards on visitor boards at conventions etc) and it wouldn't be hard to complete these for a hypothetical 6m QSO. Only the most diligent awards manager would spot the fake if the details looked reasonably plausible. There are rumours that some high scores on the 6m DXCC Honor Roll were boosted this way but one has to wonder whether the applicants concerned really get any satisfaction from such behaviour.

But also in this category there are any other forms of behaviour that are against the rules of the awards programme or contest concerned. It gets complicated. A contest sponsor may, for example, prohibit the use of Internet-based support (Cluster, ON4KST chat), but these are quite valid for most awards. So, you may decide to operate with such support in order not to miss any rare ones that appear during the contest, but not to send in a contest entry. That's absolutely fine. But as soon as you cross over the line, whether in terms of contest rules or award rules, then, again, there is surely no satisfaction to be gained because your score will not have been gained on the same playing field as other participants.

Matters of Choice

For most (hopefully all!) readers the previous couple of paragraphs received your immediate assent. But now we get on to much more grey areas, because many of them come down to personal choice. The classic example is the use of the WSJT suite of programs, which has caused a lot of controversy in recent years. Quite simply, the major awards (let's take DXCC as the prime

THE DX CODE OF CONDUCT

1) I will listen, and listen, and then listen again before calling.
2) I will only call, if I can copy the DX station properly.
3) I will not trust the DX cluster and will be sure of the DX station's callsign before calling.
4) I will not interfere with the DX station nor anyone calling and will never tune up on the DX frequency or in the QSX slot.
5) I will wait for the DX station to end a contact before I call.
6) I will always send my full callsign.
7) I will call and then listen for a reasonable interval. I will not call continuously.
8) I will not transmit when the DX operator calls another callsign, not mine.
9) I will not transmit when the DX operator queries a callsign not like mine.
10) I will not transmit when the DX station requests geographic areas other than mine.
11) When the DX operator calls me, I will not repeat my callsign unless I think he has copied it incorrectly.
12) I will be thankful if and when I do make a contact.
13) I will respect my fellow hams and conduct myself so as to earn their respect.

example) allow contacts made by data modes to be counted towards your overall totals. So, if you aspire to sit near the top of the tables you will need to include those modes in your armoury and be ready to operate meteor scatter, EME etc with all the tools at your disposal.

But for many operators such software is anathema. They consider that it is nothing more than computer talking to computer and is no measure of true operating achievement. "How many squares has your computer worked recently?" is a typical put-down remark.

There is no right or wrong answer. If you have sweated for years to achieve a high score by what might be called 'traditional' means, you will probably be frustrated to see others catching you up at a rate of knots by using these recent advances. But if, for example, you are operating from a somewhat limited urban location, you will probably regard the digimodes as a lifesaver, allowing you to explore propagation modes that, otherwise, would be quite out of your reach. And there is no doubt that the digital modes have given a spur to operation from countries where English isn't widely spoken because contacts can be made without any need to speak.

A slightly different example is the situation discussed elsewhere in this book where your station is remote from your home. Some operators would gain no satisfaction from having to resort to such means to work DX and would only be satisfied with contacts made from home. Others consider it the best way to be competitive on a band which is very sensitive to location. There is no right or wrong answer, though lines have to be drawn. Both *CQ Magazine* and the ARRL, for example, have minimal restrictions on remote operation for awards purposes but both bodies place restrictions on the location of equipment and, generally, operators for contest operations.

At the end of the day most of these things are a matter of personal choice and you need to decide what matters to you and then, on the whole, stick with it so that you can measure your personal progress in a meaningful way even if you cannot always compare your results with the published results of others.

And finally, while on the subject of ethics and good operating, it is worth pointing out that the RSGB, UKSMG, many expedition teams and others nowadays highlight their support for the 'DX Code of Conduct' (**[11]** and reproduced in the sidebar on page 109). Given that deliberate QRM, bad operating and similar seem to be on the increase (albeit perhaps more on HF than at VHF) this Code cannot be repeated often enough although, sadly, those who ignore it are probably those who would never actually be aware of it in the first place.

REFERENCES

[1] Daily DX: www.dailydx.com
[2] OPDX Bulletin: www.papays.com/opdx.html
[3] 425 DX News: www.425dxn.org
[4] DX-World: www.dx-world.net
[5] DX Summit: www.dxsummit.fi
[6] ON4KST: www.on4kst.com/chat/start.php
[7] CW Skimmer: www.dxatlas.com/CwSkimmer
[8] RBN: www.reversebeacon.net
[9] PSKReporter: https://pskreporter.info/pskmap.html
[10] UK 4m reflector: http://groups.yahoo.com/group/fourmetres
[11] DX Code of Conduct: http://dx-code.org

Chapter 6

Making Weak-Signal QSOs

THE 6M BAND IS great fun when there is a good F2 or Es opening, 4m similarly during the Sporadic E season, but much of the time that simply isn't the case. Probably for 90% or more of the time, in fact, with Es present for just a month or two each year (and, even then, only for a few hours each day most of the time) and F2 propagation (on 6m) only at the peak of the sunspot cycle.

What to do at other times? You could ignore Six and Four other than for local chats and stick to HF. Or you could stick with Six and Four and continue to work DX using 'weak signal' modes. Until almost 20 years ago (yes, WSJT arrived as long ago as 2001), the only weak signal mode available to amateurs was CW, which has a theoretical advantage of several dB over SSB, providing you use narrow filters to ensure the best signal-to-noise ratio. That is why this book, in several places, recommends that you be ready to use CW when necessary, for example to catch that DXpedition during the short opening when his signal appears above the noise. If this book had been written 20 years back, this chapter would have still been about making use of meteor scatter and EME, but the emphasis would very much have been on using CW. It would have talked about the need to squeeze every last dB out of your station to work these modes effectively, but they would have remained very much the domain of the handful of best-equipped 6m enthusiasts and largely out of reach of the majority. 4m operators could try meteor scatter, but there wouldn't have been many people to work.

Since then the world of weak signal working has changed dramatically. Nowadays there are other weak signal modes available to amateurs. HF operators will be familiar, for example, with PSK31, which has been around for many years now. It allows contacts to take place with low power when, at times, you simply can't hear the station in your headphones although you still have solid copy on the screen.

INTRODUCING WSJT

For EME working the WSJT mode JT65 came to the fore several years ago while, for meteor scatter, FSK441 and JT6M became popular. However, what has transformed DXing (and, indeed, day to day QSOs) on both 4m and 6m, is the arrival of FT8 and, more recently, FT4, again parts of the WSJT suite of weak-signal programs. And, for meteor scatter working, MSK144 is now the mode of choice.

Joe Taylor, K1JT, pictured at Visalia, California.

On VHF, weak signal operating has largely come to mean WSJT, the software suite developed by Nobel Laureate Joe Taylor, K1JT (hence the name: Weak Signal by Joe Taylor). WSJT has been available since 2001 and has been progressively developed, with new modes being added and existing ones improved. JT6M was released in 2003, with many subsequent updates, not least FT8, FT4 and JT9. No doubt that development will continue in the years to come. For the purposes of this discussion, I won't make a distinction between WSJT and WSJT-X – the 'X' extension simply refers to what started as an experimental branch of WSJT but has now become the mainstream program.

Because the WSJT-X software is in the public domain, various clones have been developed based on the underlying code, the most popular of which is JTDX. While developed primarily for HF DXing, JTDX is popular among a number of 6m operators. The interface is very similar to WSJT but the screen layout and some of the facilities are different. JTDX has been developed by Igor Chernikov, UA3DJY, and Arvo Järve, ES1JA, and supports several of the WSJT-X sub-modes, including JT9, JT65, FT8 and FT4. Some users prefer its user interface to that of WSJT-X and claim that it is better at decoding the weakest signals.

Another program derived from WSJT-X is MSHV by Christo Hristov, LZ2HV. This supports, among other sub-modes, ISCAT, JT6M, JT65, FT8 and FT4.

For simplicity, this chapter will focus primarily on WSJT, still the most widely used of the weak signal programs and, in any case, much of what follows is equally pertinent to JTDX and MSHV although you will need to consult the relevant documentation should you choose to use either of those programs. WSJT allows signals to be recovered from below the ambient noise floor and is generally considered to have a 10 – 15dB advantage over CW, a huge difference when you consider that a linear amplifier typically adds 10dB or so to your signal and an improvement of 10dB in antenna gain requires something like eight times as many elements as you already have (on the basis that doubling the antenna size typically adds 3dB or so)! The impact of that level of improvement is that WSJT brings weak signal working to a much greater number of 6m operators than previously. It also opens up

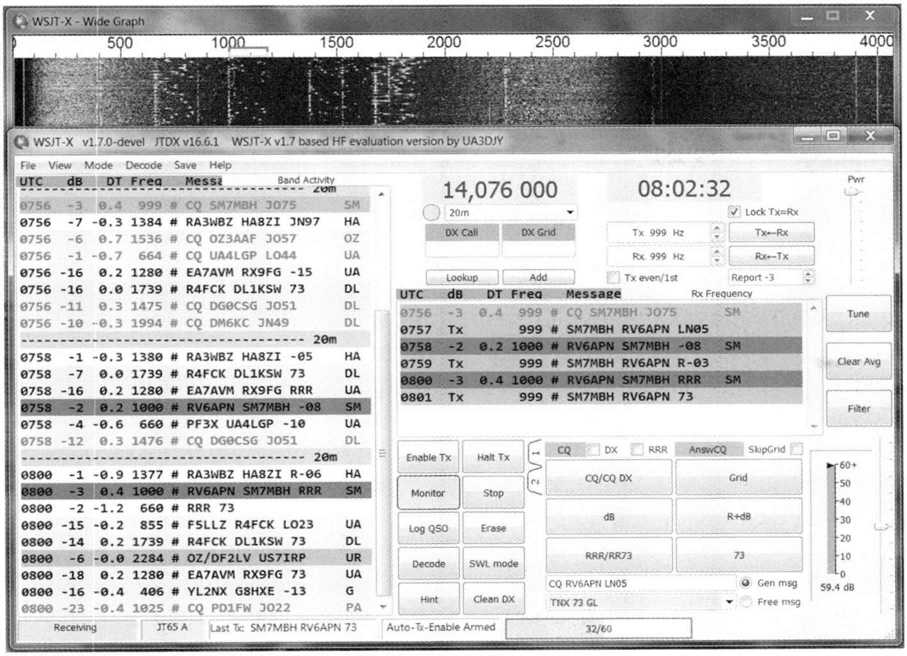

A screenshot from JTDX (taken from the JTDX website).

propagation paths that were previously thought impossible.

Several operating modes are available within WSJT, designed for different types of working. JT65A, for example, is optimised for EME working where signals are continuous but extremely weak. JT6M and ISCAT (now effectively replaced by MSK144) are optimised for 6m meteor scatter working, where signals may be stronger but are intermittent, being reflected off the fleeting ionisation trails which meteors leave as they enter the atmosphere. ISCAT affords the option of either 60 or 30 second sequences, and was designed primarily for marginal ionospheric propagation on 6m. It is not adversely affected by random meteors (as is the case with JT65A), so it can be used for paths from meteor scatter range to multihop ionospheric propagation. However, nowadays MSK144 has become the more popular mode for meteor scatter working on 6m, albeit originally developed for 2m.

JT65 and then JT65A has been used for weak-signal terrestrial working too but what has transformed 6m operating has been the arrival of FT8. While not so sensitive as JT65A, the 15 second 'overs' (as against one minute for JT65A) mean that QSOs can be completed more quickly, useful not only to maximise QSO rate but also helpful if propagation is changing rapidly.

FT4 is faster still and while it was developed primarily for contest working, it has also been pressed into service as a day-to-day QSO mode when signals are reasonably strong (because the shorter decode time makes it less sensitive than FT8).

As a result, when you are enjoying S9 contacts around your continent by Es, for example, your neighbour may be working DX several thousand kilometres away, by multiple-hop or extended Es, using WSJT to pull out signals that are inaudible to the ear. So, while 6m and 4m operators now routinely use FT8 and FT4, these modes really come into their own to work DX that SSB and CW operators simply cannot hear.

At the moment JT65A remains the most sensitive mode within WSJT, so it is still the standard used for all 6m EME. Because of its great sensitivity, it also has been used during marginal multihop Es and F2 propagation but has pretty much fallen into disuse for this purpose since the advent of FT8. It also can greatly extend the range of contacts on ground wave or D-layer scatter propagation, provided there are no meteors to disrupt the decoding. If used over distances less than 1300 miles, it is usually recommended to avoid meteor showers or mornings, when random meteors are most prevalent.

For example, W7GJ reports having completed D-layer contacts in the afternoons with stations 600 to 1300 miles away using JT65A mode while both stations were running 100 watts with single Yagis. A contact range that used to be possible only between very large, well equipped stations is now available to very modest sized stations using JT65A. (But, again, FT8 has pretty much replaced JT65A because it is almost as effective and is now so widely used.)

Of course, MSK144 mode is so effective on random meteors that many modest 6m stations can use that mode for contacts under 1300 miles if they wisely choose the proper time of day to best take advantage of the meteors. And, as mentioned, FT8 makes possible daily QSOs over distances much greater than line of sight.

Some amateurs decry WSJT and other digital modes as not being 'proper' amateur radio, in that the PC appears to be doing the work, with minimal intervention from the operator. At the end of the day no-one is going to force you to use WSJT. But it can and does allow 6m enthusiasts to continue enjoying long-distance contacts when more traditional modes aren't up to the job. Which, surely, must be a good thing, if only to maintain band occupancy?

From an awards point of view, WSJT contacts are just as valid for most awards (including DXCC) as contacts made via any other mode. Choosing *not* to use FT8 will certainly put you at a disadvantage in the DX stakes!

WSJT requires accurate timing of transmissions and EME contacts are often made by way of a schedule ('sked'), to ensure that both parties are on the same frequency and that the timing of their transmissions is co-ordinated. In the past this may have been organised via the telephone or maybe a QSO on HF.

Nowadays it will almost certainly be made via the Internet, for example via the ON4KST chat page, which has been mentioned elsewhere. This is fine and accepted, but no QSO information (signal reports etc) should be passed by those means or the contact becomes invalid. Common sense needs to prevail here. If you and your QSO partner are experimenting, a high level of chat via the Internet may be appropriate to compare notes on how things are going and what, if anything, is being copied. But if you are chasing a contact which you may later need as part of an awards claim, you need to ensure that no accusation can be levelled against you that the contact was in some way compromised by the exchange of key information via the Internet or other extraneous means.

SO JUST WHAT IS WSJT?

Developed by K1JT for weak signal working (not strictly true, as mentioned above, as meteor scatter is not necessarily weak signal, but characterised by short 'pings' or bursts of signal between periods of silence) on the VHF bands, the description in the manual says: *"WSJT is a computer program for VHF / UHF communication using state-of-the-art digital techniques. It can decode signals propagated by fraction-of-a-second reflections from meteor trails, as well as steady signals more than 10dB weaker than those needed for conventional CW or SSB."*

WSJT is Windows based, and will run on any modern PC. Linux, FreeBSD and Macintosh OS/X downloads are also available and it is worth noting that it runs well on Raspberry Pi. In technical terms, it uses the processing power of a modern PC to integrate weak signals over time in order to be able to extract them from the ambient noise, something which the human ear is not geared up to do.

The program interfaces to most modern radios and is able to output an ADIF file for upload to Logbook of the World or to import into your regular station logging program.

For 6m enthusiasts the facilities offered by WSJT are as follows:

- JT65A, a weak signal mode, designed for EME and extreme troposcatter, and replacing the previous JT44.
- MSK144, designed primarily for 144MHz meteor scatter, but now in regular use on both 50MHz and 70MHz.
- MAP65, an addition to JT65 for EME working, designed to merge signals received with orthogonal polarisations to minimise the effects of Faraday rotation. Useful too for giving a wider picture of activity as being heard at your station.

A detailed description of the WSJT program suite is provided in the online WSJT user guide. The starting point is obviously the main WSJT site of K1JT **[1]** from which you can download both the software itself and the manual. The website is also a valuable resource for would-be WSJT operators, with much advice on setting up WSJT. And while intended primarily for HF operators, the FT8 guide by Gary, ZL2IFB **[2]**, is excellent and well worth a read before embarking on FT8 operation. For an explanation of WSJT, a good place to go is the pair of articles written by K1JT, K9AN and G4WJS in the October and November 2017 issues of *QST* **[3]**. In the second of these the authors state: "FT8 is designed specifically for propagation conditions such as multi-hop sporadic-E at 50MHz – circumstances where signals are weak and fading, openings short, and quick completion of reliable, confirmable contacts is particularly desirable." A detailed technical explanation of the FT4 and FT8 communication protocols appeared in the July / August 2020 issue of *QEX* and is also available online **[4]**. Similarly, a description of the MSK144 Protocol was published in the July / August 2017 issue of *QEX* **[5]**.

But just to give a flavour of the improvement achieved with the use of WSJT, here are some typical weak signal limits by mode (signal-to-noise ratio in a 2500Hz bandwidth):

SSB	10dB approx
MSK144	–8dB
CW (by ear)	–15dB
FT8	–21dB
JT4	–23dB
JT65	–25dB
JT9	–27dB
QRA64	–27dB
WSPR	–31dB

SETTING UP YOUR STATION FOR WSJT

Your existing 6m station will almost certainly be adequate for meteor scatter operation using MSK144. Indeed, too 'sharp' an array often misses signals coming into the broader capture angle of a smaller single Yagi. Experienced operators tend to favour two or more stacked Yagis since these give extra gain while retaining a broader E-plane lobe. But many successful QSOs have been made using less than 50 watts and a 3-element Yagi.

Do remember, when using WSJT, to turn off speech processing and disconnect your microphone when using digital modes, and keep your power level down (or install additional cooling) as most amateur transceivers are rated for intermittent duty cycle (SSB and CW) whereas data modes have 100% duty cycle when on transmit.

For FT8 and FT4, again a simple station will allow you to make plenty of near-in (one hop) QSOs but also a good number of multiple-hop QSOs. However, it's true to say that serious terrestrial DXing on 6m requires a more substantial station, by which I mean at least, say, a 5-element Yagi at maybe 50ft and as much power as your licence allows.

Frequency setting is of paramount importance. A tolerance of ±500Hz is demanded on CW and 200Hz on SSB. WSJT also requires a high degree of accuracy. A stability of better than 100Hz/h is also expected. Most modern transceivers should be up to the task.

The timing requirements are stringent but not too difficult to meet. As a guide, your PC clock needs to be accurate to much less than a second. There are several software programs that can be used to ensure that the computer's clock is always spot on. Software such as Meinberg NTP or Dimension 4 **[6]** is suitable. Windows XP time synchronisation is insufficient as it checks the time only once every 24 hours, and your PC clock may drift beyond what is acceptable during that time. Later versions of Windows appear to offer adequate time synchronisation. You can also customise time synchronisation through adjustment of the Windows registry.

Although slower computers have been shown to work, the recommended minimum computer speed is 800MHz. The computer will also need a serial port or a USB-to-Serial Port Converter, in order to provide automatic computer keying of the transmitter's PTT line. An interface to connect the radio to the computer is typically required. As long as the interface permits operation of PSK31 or RTTY mode while successfully keying an amplifier, it should work with WSJT.

There are no special requirements as far as the transceiver is concerned, except insofar as it is handy to be able to monitor both 50.313MHz and 50.323MHz simultaneously, which is feasible with those transceivers having two receivers. It is helpful if the receiver has an effective noise blanker and the AGC can be turned off (though it is usual to turn off the noise blanker and noise reduction as well). It is also essential that the receiver has the capability of a wide (at least 2.8kHz and preferably as much as 5kHz) filter position for USB.

The bandwidth requirements vary with the different modes. MSK144, for example, requires a filter of at least 2.6kHz bandwidth, as do FT8 and FT4. JT65A can be used with a 1kHz bandwidth as long as Doppler or your QSO partner's possible transmit frequency error does not push the received signal out of this bandwidth.

It is not important to have DSP capability in the receiver, since the WSJT software is a digital signal processing package in its own right, working in conjunction with your PC's sound card. As such, it is important when using WSJT to have other sound sources within the computer turned off (e.g. message alerts, music accompanying web pages etc) or these sounds will end up on air. Always remember to disconnect the microphone, too. The problem with any DSP noise reduction is that it can remove the valuable information encoded into the digital messages. Therefore, you should always make sure to turn off any DSP noise blankers or noise reducers. Unless you have 'birdies' or other interfering adjacent signals, it is advisable to keep your USB bandwidth as wide as possible on your particular receiver. Using narrower filter settings introduces loss, and it is always best to let the WSJT program filter out the weak signals at audio level rather than having you introduce attenuation before that point.

While meteor scatter can be achieved with modest power levels, in order to ensure two-way contacts off the moon it is necessary in most cases to use an amplifier. A 400-watt amplifier should be the minimum considered and, typically, a kilowatt amplifier is used. Suitable models have already been mentioned in Chapter 2, but whenever you use a digital mode, and WSJT is no exception, bear in mind that the duty cycle is extremely high because the signal consists of continuous multiple tones. Many amateur radio amplifiers are rated for SSB duty so should be de-rated by 50% or more for digital modes. So, for example, to run the maximum allowable power in the UK, 400 watts at the antenna, you are probably looking at needing an amplifier sold as being capable of 1kW output. Even then, you would be best advised to consult the supplier on its suitability for extended digital transmissions.

This is perhaps the correct point at which to mention that under the Decode tag in WSJT there is the option to select Fast, Normal or Deep. Deep decoding will pull out the weakest signals but makes the greatest demand on the processor. If your PC is not up to it, using deep decoding can result in a lag while the PC catches up after the 15 second receive period. Of course, if your PC is sufficiently powerful and you are not trying to run multiple instances of the software or of other programs, there should be no problem.

Mention should also be made of the benefits of using 24-bit sound cards for decoding. Most PCs use 16-bit sound cards but the experience of serious WSJT users is that using an external 24-bit sound card (such as that which comes as part of the Microham MkIII keyer or is available in several guises for music enthusiasts such as the now obsolete E-MU 0204) can improve decoding of the weakest signals by quite a considerable margin. Similarly, regular weak-signal chasers recommend increasing the audio level within WSJT to well above that generally recommended in the manual (to perhaps 70 – 80% on the WSJT display).

GETTING STARTED WITH WSJT

This is not the place to reinvent the wheel: the WSJT user guide and online help will tell you what you need to know and the websites and other sources previously mentioned contain a lot of useful hints and tips from those who have been there and done it. The exact setup for FT8 and FT4 (indeed, for WSJT generally, or JTDX if you prefer that program) is a matter for your transceiver, your PC and the program that you choose to use, so it is impossible to give an explanation here.

Familiarise yourself with the software, test your equipment and practice making dummy QSOs by linking up with others via, for example, the ON4KST chat page **[7]**. It is important to familiarise yourself with the QSO procedure as it is easy to make mistakes in the heat of the moment when you are trying to make your first WSJT contacts.

Probably the best advice is to watch WSJT QSOs being made (there are plenty of FT8 and FT4 QSOs to be seen on 6m when the band is open, or just check any of the HF bands).

The rest of this chapter deals mainly with FT8 and FT4 operation, although meteor scatter and EME operating are covered too. Remember that while FT8 and FT4 activity is confined mainly to those times of the year when Sporadic E is in evidence, meteor scatter and EME QSOs are feasible throughout the year. Of course, if and when F2 propagation returns, that will probably peak in the autumn and spring seasons.

CONFIGURING WSJT-X FOR FT8

If you're new to data modes, I suggest you start by using VOX for transmit/receive switching rather than full CAT control. This makes the configuration simpler and you can add CAT control later. Assuming you now have WSJT-X installed, the first task is to enter your station details as follows:

1. Run WSJT-X
2. Select File - Settings and then the General tab
3. Enter your callsign and Maidenhead locator
4. Select the Radio tab and make sure the Rig is set to None and PTT set to VOX
5. Select the Audio tab and choose the soundcard you will be using for the audio input and output from your rig
6. Click OK to finish and save the configuration

When WSJT-X started you will have noticed that two windows opened, one of which was the Wide Graph. This displays the audio spectrum currently being received via the selected soundcard.

In addition to providing a view of band activity, we can use the spectrum controls to set our transmit and receive frequencies. However, to get the best display for FT8 operations we need to make some minor adjustments as follows:

1. Connect-up your audio link between the rig and computer
2. Tune your rig to, for example, the 14MHz band and select USB or Data.
3. Start WSJT-X and set the mode to FT8
4. If all is well, you should see signs of life in the Wide Graph and the level meter in the bottom left of the main screen.
5. Tune your rig to an empty RF channel somewhere around 14.1MHz
6. Adjust the audio input level to the computer to get a reading of approximately –30dB in WSJT-X
7. Retune to the main FT8 frequency of 14.074MHz
8. You should see some FT8 activity as this band is rarely quiet
9. Make sure the input level bargraph remains green

With the receive sensitivity configured, we need to adjust the Wide Graph parameters to optimise the view for FT8 as follows:

1. Select the Wide Graph window
2. Set the Bins/Pixels to 5, Start to 100Hz and N Avg to 2

The next step is to adjust the waterfall and spectrum sensitivity sliders. These are located on the lower-right of the Wide Graph and the top two sliders adjust the waterfall while the lower set deal with the spectrum display.

Having completed the receive setup, we can move on to the transmit configuration. One common failing is to overdrive the audio stages of the transmitter. This results in the transmission of multiple distorted artefacts of your signal, so you won't be popular!

To configure the transmit path, begin with a dummy load connected in place of the antenna. You can then click the Tune button in WSJT-X to generate a transmit audio tone. In the bottom right of the WSJT-X main panel you will find the transmit drive slider. This slider adjusts the level of the transmit tone and I suggest starting with this set to about 75%. Now you can move to your rig and adjust the mic audio gain, VOX sensitivity and the PA drive to reach the target output power. Start with maybe 5 watts. If you have a digital scope to hand, you could put it in spectrum analyser mode and use a tap of the transmit output to check the spectrum quality.

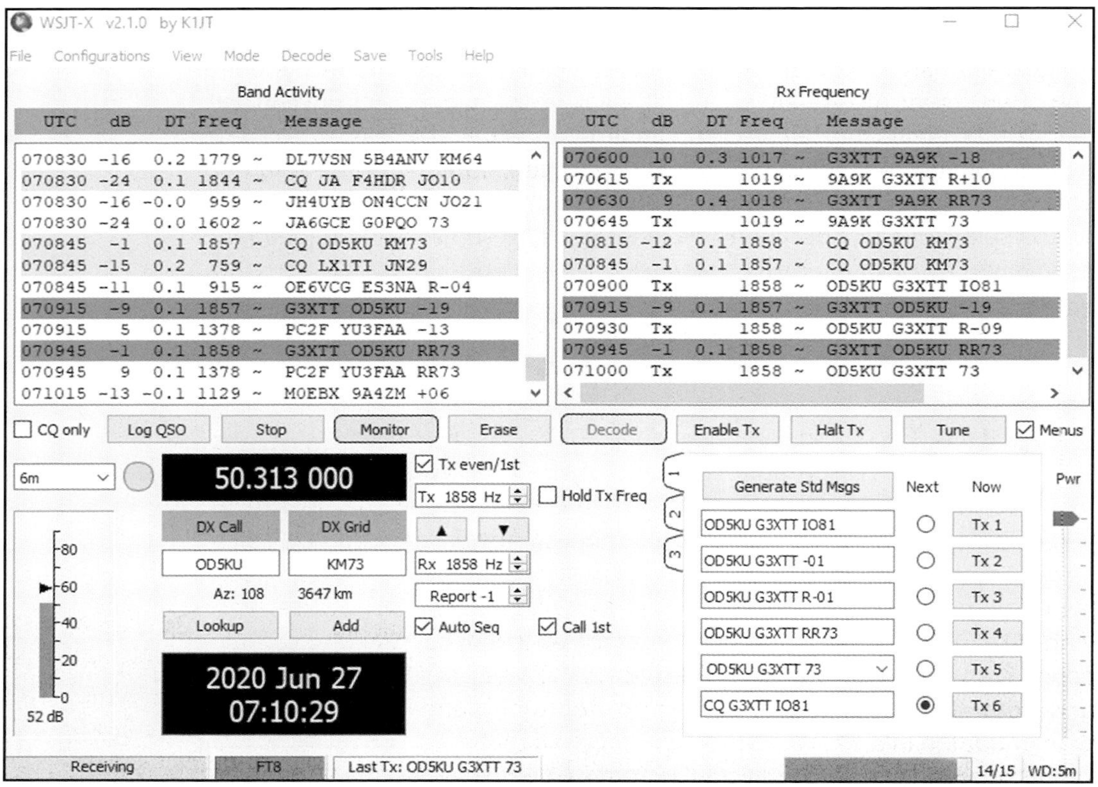

A screenshot from the WSJT software, running FT8 on 6m.

When you start making QSOs, you use the transmit slider in WSJT-X to trim the power output.

TUNING FT8

Because FT8 signals use such a low data rate, the decoding process requires very little processing power from your PC. FT8 software takes advantage of this and can comfortably decode a large number FT8 signals in each 15-second time slot. As a result, it is standard practice to confine FT8 operations to a single, SSB width, channel in each band. This greatly simplifies operating because you can tune your rig to a single SSB frequency and monitor and communicate with any active station without touching the rig's tuning.

Another important characteristic of FT8 operation is the use of split transmit and receive frequencies. As the decoder can handle all the signals in the 2.5kHz bandwidth, you can reply to a station using any clear frequency.

The first task before making a call is to select a clear transmit frequency. You can find a suitable frequency by watching a few 15-second FT8 cycles on the Wide Graph display. Once you've identified a clear frequency, use the mouse to shift-click on the left-hand edge of the chosen frequency. You will see the red, transmit frequency, marker move to that location. When you've set the transmit frequency, make sure you tick the Hold Tx Freq box in the lower centre. This will lock the transmitter on that frequency, leaving you free to move the Rx frequency to the station you want to work.

WSJT-X does a particularly good job of simplifying FT8 operation. The operational heart of the software is the two main panels. The left panel shows all the decoded activity, while the right-hand panel shows the activity on the currently selected receive frequency (green marker on Wide Graph). CQ calls will be highlighted in the band activity panel and you can respond to a CQ by double-clicking on that station's entry. This will populate the call and locator fields and set the transmitter to start on the opposite cycle to that of the wanted station. Once the link has been established, the software will automatically step through the structured messages and complete the QSO.

FT8 AND FT4

FT8 transmissions are generated using audio tones. In the case of FT8, eight-tone Gaussian Frequency Shift Keying (8-FSK) is used with a slow baud rate of just 6.25 bauds. This modulation systems employs eight, closely spaced, frequencies and the gaussian filtering ensures a smooth transition between tones and reduces the key-clicks that would otherwise be generated by abrupt frequency changes. By using such a slow baud rate, FT8 signals occupy a very narrow bandwidth of just 50Hz. The use of such a narrow bandwidth means the entire FT8 activity for each band can normally be accommodated in a single 2.5kHz speech channel.

The operating procedures for FT8 and FT4 have their origins in the techniques developed for EME (Earth-Moon-Earth). When working EME,

FT8 INTERCONTINENTAL DX CODE OF PRACTICE FOR 50.323MHZ

During the Northern Hemisphere Summer Es season the use of 50.313MHz for FT8 contacts has become very congested, and stations are now using 50.323MHz for Intercontinental DX working. In order to not cause QRM to each other and permit the most efficient use of FT8 for weak signal DX working, it is highly desirable for all stations to operate in the same transmit windows. The UKSMG has prepared these guidelines for the use of 50.323MHz:

1. European stations to transmit in the "even" periods. That is 00, and 30 seconds past each minute.
2. Asian and North American stations to transmit in the odd periods. That is 15, and 45 seconds past each minute.
3. Do not use 50.323MHz for working stations within your continent. Europeans working other Europeans are strongly discouraged.
4. Familiarise yourself with the excellent FT8 operating guide by ZL2IFB at http://www.g4ifb.com/FT8_Hinson_tips_for_HF_DXers.pdf

(Please note that this code of practice is a work in progress.)

the exchange required for a completed QSO comprises the callsign, report and other information such as the location. In addition to a fixed format for the messages, EME operators use timed transmit and receive slots to improve their chances of success. FT8 follows similar principles by using 15-second timeslots and structured messages. Other than a 13-character (including spaces) final message, there is no facility in FT8 for free text messages.

Rather than use the usual RST report, the FT8 software automatically calculates the signal-to-noise ratio based on a 2.5kHz bandwidth. One of the benefits of using the WSJT software is that the whole process is automated, albeit the authors insist on a minimum of operator intervention by way of logging the QSO and manually restarting the next one. But to be most effective, especially when DXing, quite a bit of operator intervention is usually required.

For example, although the software offers the option of co-channel operation (going back to the calling station on his own frequency), this is rarely to be recommended when chasing DX because if the DX station has several callers on his frequency, the software will be unable to decode them. So, turn on 'hold tx freq' and use shift-click to select a clear spot on which to call. And if, for example, you are using PSKReporter, you can hopefully see your calls being spotted. If not, that could mean the frequency you have chosen is actually busy at the other end and you may need to QSY.

The following explanation of a standard exchange is from the WSJT Manual:

Standard Exchange

By longstanding tradition, a minimally valid QSO requires the exchange of callsigns, a signal report or some other information, and acknowledgments. WSJT-X is designed to facilitate making such minimal QSOs using short, structured messages. The process works best if you use these formats and follow standard operating practices. The recommended basic QSO goes something like this:

CQ K1ABC FN42	(K1ABC calls CQ)
K1ABC G0XYZ IO91	(G0XYZ answers)
G0XYZ K1ABC –19	(K1ABC sends report)
K1ABC G0XYZ R –22	(G0XYZ sends R+report)
G0XYZ K1ABC RRR	(K1ABC sends RRR)
K1ABC G0XYZ 73	(G0XYZ sends 73)

Standard messages consist of two callsigns (or CQ, QRZ, or DE and one callsign) followed by the transmitting station's grid locator, a signal report, R plus a signal report, or the final acknowledgements RRR or 73. These messages are compressed and encoded in a highly efficient and reliable way. In uncompressed form (as displayed on-screen) they may contain as many as 22 characters. Some operators prefer to send RR73 rather than RRR. This is workable because RR73 is encoded as a valid grid locator, one unlikely ever to be occupied by an amateur station.

Signal reports are specified as signal-to-noise ratio (S/N) in dB, using a standard reference noise bandwidth of 2500Hz. Thus, in the example message above, K1ABC is telling G0XYZ that his signal is 19dB below the noise power in bandwidth 2500Hz. In the message at 0004, G0XYZ acknowledges receipt of that report and responds with a –22dB signal report. JT65 reports are constrained to lie in the range –30 to –1dB, and values are significantly compressed above about –10 dB. JT9 supports the extended range –50 to +49dB and assigns more reliable numbers to relatively strong signals.

And as mentioned already, always tick the 'Hold Tx Freq' in WSJT to ensure that you are not transmitting on top of the called station. This is good practice with any QSO but especially when calling a DX station. It can be thought of similarly to operating 'split' on SSB and CW.

And European stations must remember to tick 'TX even/1st' when working the USA, again to avoid unnecessary interference on the weakest of DX signals.

It's also worth going into File / Settings / Radio and under Split Operation click 'Fake it'. What this does is to make sure that wherever you are working someone on the waterfall, at the top or the bottom, it will always transmit at a tone frequency of around 1500Hz (so if you're low in the waterfall, the transmit frequency might be 50.312 (say) and if you're very high in the waterfall maybe 50.3135. The benefit of this is that any nasty audio harmonics

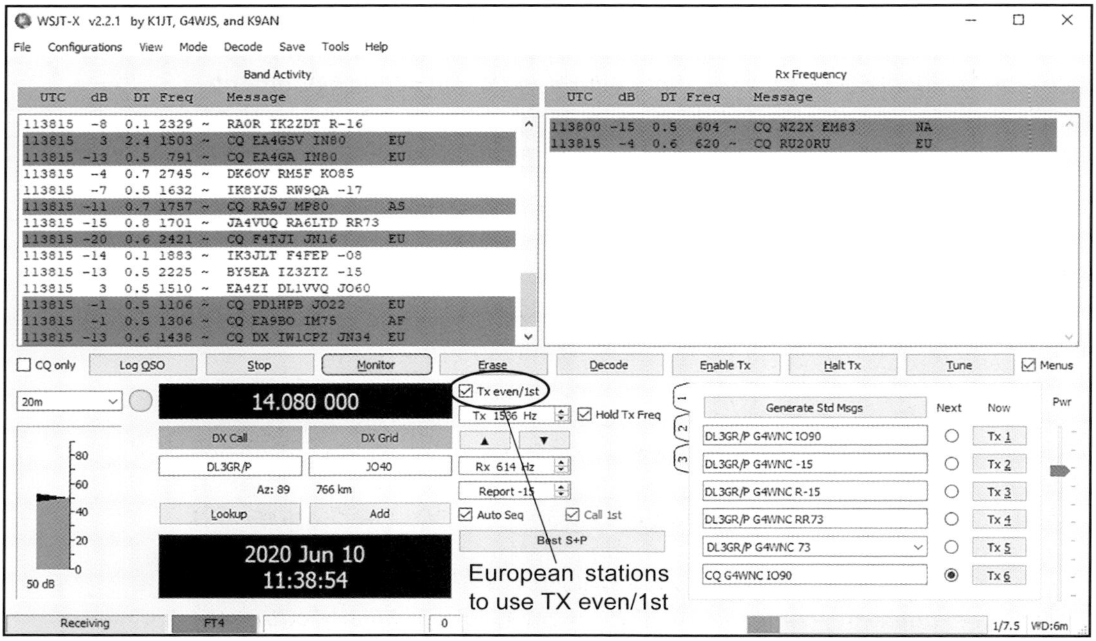

European stations be sure to tick 'Tx even/1st'.

you might have been transmitting will be stopped by the filter characteristics of the radio.

FT8 DXPEDITION MODE

A brief mention of FT8 DXpedition mode. DXpedition mode is intended primarily for rare-entity DXpeditions on the HF bands and you are unlikely to encounter it on 6m but it is entirely possible that some DXpeditions may use it on 6m in the years to come. This special operating mode enables DXpeditions to make FT8 QSOs at very high rates by allowing the DXpedition to respond to multiple stations at the same time. The setup and operating procedures differ from normal FT8 QSOs and should be made in accordance with the FT8 DXpedition Mode Manual (from the WSJT website).

FT4

FT4 was developed as a contest mode, with a minimal contest exchange and less effective at digging down into the noise than FT8 but allowing a much quicker QSO. But these advantages have been taken on board for day-to-day DXing too, where signals are strong enough not to need the deep decoding of FT8. It will be interesting to see how the use of FT4 on 6m and 4m develops over time but here is the gist of recent post about Joe's, K1JT, use of FT4 on the 6m band:

Joe, noting the transatlantic six meter opening(s), says he is enjoying the multi-hop sporadic E, "an excellent test of the original design goals for

FT8, and more recently also FT4." Over several hours he worked 76 European stations, 37 with FT8 on 50.313 (mostly) or 50.323. The other 39 were FT4 on 50.318, "in just 44 minutes." Joe says, "I strongly recommend that for such openings in the future you should try FT4. It's twice as fast as FT8, with only a 3db penalty on the AWGN channel. The sensitivity penalty is even less on a fading channel."

Joe says it was also an excellent opportunity to use good operating practices. "As recommended, most European stations transmitted in the first / even sequence in both FT4 and FT8. North American stations transmitted in second sequence. This procedure minimizes QRN from strong local stations and helps everyone to work DX."

METEOR SCATTER

Meteor Scatter (MS) contacts are made by reflecting signals from the ionised trails of meteors during atmospheric entry (see **Fig 6.1**). MS operation is feasible on both 50MHz and 70MHz (also 28 and 144MHz, but that is outside the remit of this book).

Commercial MS links tend to use the low VHF region, as the path availability is greater than at the higher frequencies. The distances which may be covered by typical MS operation are similar to those possible via Sporadic E. Assuming most signals are reflected from a region at an altitude of 110km, the maximum range possible with an antenna exhibiting a main lobe at 0° elevation is about 2300km.

Typical amateur antennas have a main lobe at 2 – 5° and thus the ranges to be expected are somewhat less. Most MS contacts on 2m take place during periods of more-or-less predictable meteor activity – the so-called *meteor showers*. Rather fewer contacts are made via sporadic meteors during the intervals between. In contrast, 6m meteor scatter activity is possible on

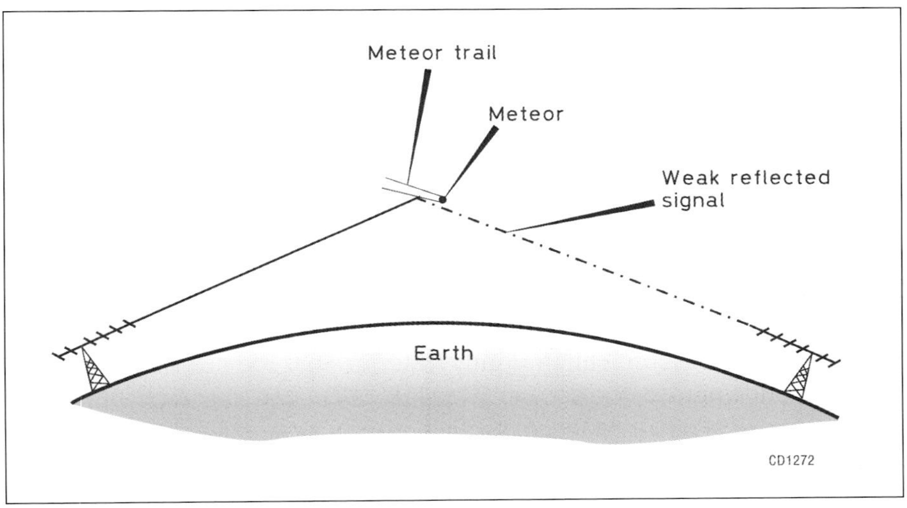

Fig 6.1: How VHF signals can be reflected by a meteor trail.

Shower name	Limits	Max	ZHR	N–S	NE–SW	E–W	SE–NW
Quadrantids	1–5 Jan	3–4 Jan	110	02–06(W) 11–16(E)	11–17(SE)	23–03(S)	00–05(SW) 15–17(S)
April Lyrids	19–25 Apr	22 Apr	15–25	22–02(W) 06–10(E)	23–03(NW) 08–11(SE)	03–06(N)	22–01(SW) 05–08(NE)
Eta Aquarids	1–12 May	3 May	50	03–04(W) 10–11(E)	04–09(NW)	05–11(N)	08–12(NE)
Arietids	30 May–18 Jun	7 Jun	60	04–08(W) 11–15(E)	05–09(NW) 14–16(SE)	08–12(N)	04–06(SW) 10–14(NE)
Zeta Perseids	1–16 Jun	9 Jun	40	05–10(W) 13–17(E)	06–11(NW) 15–17(SE)	09–14(N)	07–07(SW) 11–15(NE)
Perseids	20 Jul–18 Aug	12 Aug	95	23–04(W) 09–13(E)	08–17(SE)	11–01(S)	18–04(SW)
Orionids	16–27 Oct	22 Oct	25	00–03(W) 07–09(E)	00–04(NW)	03–06(N)	05–08(NE)
Taurids S	10 Oct–5 Dec	3 Nov	25	02–05(E) 20–22(W)	20–01(NW)	22–03(N)	00–05(NE)
Geminids	7–15 Dec	13–14 Dec	110	04–09(E) 20–01(W)	22–02(NW) 05–09(SE)	01–04(N) 03–07(S)	03–07(NE) 19–23(SW)
Ursids	17–24 Dec	22 Dec	15	—	07–01(SE)	00–24(S)	16–09(SW)

Table 6.1: The principal meteor showers.

most days of the year using the specialist MSK144 mode. SSB and CW contacts via meteor scatter are rare nowadays as MSK144 makes life so much easier. Indeed, MSK144 (and JT6M before it) has transformed meteor scatter operation on 6m in just a few years, with activity taking place most days.

MSK144 was introduced to the WSJT software suite in January 2017, replacing JT6M and FSK441, although some users still prefer FSK441. But MSK144 offers shorter transmit and receive periods than the latter along with auto-sequencing that reduces human error and improves operator efficiency.

Table 6.1 lists the major meteor showers, which occur on the same dates each year as the earth passes through those bands of particles on its passage round the sun. Zenith Hourly Rate (ZHR) is the number of meteors per hour at the peak of the shower. This is a calculated maximum number of meteors an ideal observer would see in perfectly clear skies with the shower radiant overhead, based on historical data. For the radio amateur, it gives an idea of which showers are the most prolific.

The geographic columns show the QSO path and optimum times. For example, a UK station wanting to work Italy would look at the N-S column to find the best times for a QSO. The letters in brackets show the radiant (direction of arrival). So that, for example, for a N-S QSO during the Quadrantids, the table shows 02 – 06hrs (W) the radiant being to the west, then from 11 – 16hrs the radiant is to the east. Both times would be good as the meteor shower is crossing the N-S path.

Serious meteor scatter enthusiasts ensure they are ready for these events, with no distractions in their diaries! Nowadays, though, more casual operators may become aware of meteor scatter opportunities as they watch the DX Cluster and see stations exchanging sked information for MS QSOs. This is an example where VHF use of the Cluster system differs from HF use. While on HF it is deemed 'unsporting' to arrange skeds or self-spot via the Cluster system, on VHF the Cluster becomes a useful tool for setting up advance schedules on specialist modes. Of course, for the contacts to be valid no exchanges can be made on the Internet during the contact.

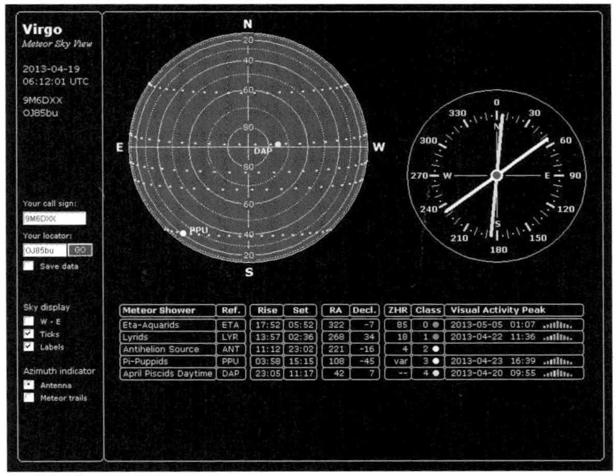

Fig 6.2: DL1DBC's Virgo meteor sky view, a projection of the 'meteor sky' at your location. The view is updated each minute using International Meteor Organization data.

A reread of the meteor scatter paragraphs in Chapter 4 would be useful at this point. As discussed there, the best way of working via a recognised meteor shower is to download data about the shower and to beam at 90° to the radiant.

Other stations will also be active during these showers and a CQ on MSK144 on 50.280MHz may well produce callers. Alternatively, arrange a sked either beforehand or in real-time via the ON4KST chat room or the Ping Jockey **[8]** websites. General information about forthcoming meteor showers can be obtained from the International Meteor Organisation website **[9]**, but perhaps more useful to amateurs is DL1DBC's meteor scatter site **[10]** – see the screen shot in **Fig 6.2** – which gives radiant and other data so that you know where to point your beam.

OPERATING PROCEDURES FOR METEOR SCATTER QSOs

While detailed operating procedures were in place previously for meteor scatter QSOs, nowadays it's pretty much a case of simply allowing the MSK144 software to take care of the QSO protocols. The WSJT manual says:

Meteor scatter QSOs can be made any time on the VHF bands at distances up to about 2100km (1300 miles). Completing a QSO takes longer in the evening than in the morning, longer at higher frequencies, and longer at distances close to the upper limit. But with patience, 100W or more, and a single Yagi it can usually be done.

Unlike other WSJT-X modes, the MSK144 decoder operates in real time during the reception sequence. Decoded messages will appear on your screen almost as soon as you hear them.

A *QEX* article describes the method for split frequency operation, for when the regular meteor scatter frequency becomes too busy:

Split Operation Using "CQ nnn"

The sporadic nature of meteor pings makes it possible for many stations to share a frequency with little interference. Decoding several different stations in the same receiving sequence lets you see "who else is on", in addition to the station you may be working. However, when the frequency gets too busy it's a good idea to spread out.

WSJT-X provides a mechanism for doing this while preserving the important advantage of having a known "meeting place" for initiating contacts. CQ messages may include three digits between the CQ and callsign. For example, K9AN might send "CQ 290 K9AN EN50" on 50.280MHz to indicate that he will listen for replies on 50.290MHz, and continue the QSO there. Note that offset CQs are not the same as typical "split" operation on HF bands. When operating split, each operator transmits on one frequency and listens on another. MSK144 QSOs using "CQ nnn" are used to move a contact off the calling frequency to a single offset frequency where both stations will transmit and receive. WSJT-X has facilities that allow it to recognise "CQ nnn" calls and reset the transceiver's dial frequency automatically, as required for both stations.

A little forethought is necessary when planning meteor scatter activity. For example, if you are in the UK, you need to catch the meteor showers when they are to the east, as that is where activity will be. There would be opportunities to work to the west later in the day, but there's generally not too much activity from somewhere out in the Atlantic Ocean! It's not the meteor shower that's moving, of course, it's the earth rotating through the meteor belt.

But many 6m and 4m meteor scatter operators don't bother with any of that. Instead they rely on the daily occurrence of random meteors entering the earth's atmosphere. As described in Chapter 4, the peak occurrence of such meteors is around dawn local time, so UK amateurs tend to focus their activity early in the day, perhaps operating from an hour or so before dawn until an hour or so afterwards. Again, most of the activity, for UK-based stations, will be to the east, where it is already later in the day and you are looking for reflections off debris ('rocks' as meteor scatter operators refer to them, despite their microscopic size!) half way between you and the station you are trying to work.

With this sort of activity, your beam heading will be towards the station you are trying to work. From other parts of Europe, or within, say, North America, the same logic applies, but you need to work out what this means in terms of whether you are wanting to work to the east or west of your QTH (or, indeed, north or south, in which case local time is the same at both locations).

Meteor scatter operating is very different from a 'traditional' QSO. The QSO is pieced together over a period of time, although 6m and 4m contacts can generally be completed much more quickly than is the case on, say, 2m. With WSJT it is possible for an entire contact to be completed during a single 100ms burst, not uncommon in the morning even when there is no meteor shower.

On SSB it should be possible to complete a meteor scatter contact within a maximum of about 15 minutes. But the main point is that only once all the data is in place can the contact be considered complete. Thus, some very specific guidelines have evolved, which are very different from what you will have encountered on other modes. They are endorsed and promoted by the IARU to ensure that all operators follow the same procedure. The purpose is to avoid confusion and to ensure absolute clarity by both QSO partners as to whether a valid contact has taken place. The Internet is there for chatting before or afterwards; it should not be used to fill in any missing data! However, the simple fact is that SSB and CW meteor scatter QSOs are essentially things of the past nowadays, WSJT having become the *de facto* way of making meteor scatter QSOs.

MOONBOUNCE

(This section compiled with substantial help from Lance, W7GJ, one of the leading operators on 6m EME and has been updated with assistance from Chris, G3WOS, who is a leading UK proponent of 6m EME.)

Probably the most challenging type of VHF / UHF DX communication is that of using the moon as a passive reflector, popularly known as EME (earth-moon-earth), or 'moonbounce'. The first known use of the moon as a relay was by the United States Navy which set up a circuit between Washington DC and Hawaii using 400MW (400,000kW!) ERP although the excellent Wikipedia entry on Moonbounce **[11]** suggests that the idea was first mooted by W J Bray of the British Post Office in 1940.

The first amateur EME contact was on 1296MHz in July 1960 between W6HB and W1BU. Back then CW was the main operating mode and, until the advent of WSJT software, EME on 6m was accessible only to those with very large stations (plenty of power and multiple, stacked antennas).

In his VHF column in the January 2000 issue of *CQ Magazine*, Joe Lynch, N6CL, wrote, "As the F2 propagation waned, those of us perilously close to the magic number 100 started to scramble to see how else we could make those last few contacts to reach the goal. For a while EME was the answer. However, that interest quickly waned as more and more of us found out just

Lance Collister, W7GJ, is one of the world's leading 6m EME specialists.

how much of an investment it took to make those precious few contacts. Many of us found that the cost per QSO, whether monetary or time, was way beyond our means. This form of propagation has all but died out for the time being on this band."

In the three decades since the first 50MHz amateur moonbounce contact between the teams of Dick Allen, W5SXD, and Joe Muscanere, WA5HNK (now W5HNK), in Texas, and Connie Marshall, K5WVX (now K5CM), and Sam Whitley, W5WAX (now K5SW), in Oklahoma, the playing field has changed dramatically. That historic contact utilised a pair of 8-Yagi arrays, each aimed at the horizon.

Previously only an option open to a very few of the world's largest stations, moonbounce contacts are now within reach for most well-equipped 6m stations. Until recently, an accomplishment such as making a moonbounce contact on 6m EME was about an S-unit or two beyond the reach of most stations. And when it comes to something like 6m moonbounce, where signals are just on the threshold of detectability, 5 to 10dB is a huge amount! However, several very significant developments have combined to make this out-of-this-world type of 6m propagation practical today.

6m is indeed the most challenging band in the VHF / UHF spectrum due to the physically large size of 6m antennas, the high background sky noise, the many geomagnetic disturbances that can deflect the moon-bound and / or earth-bound signals off course, and the prevalence of local noise around 50MHz. But what has changed since 2001 is the arrival of WSJT. By taking advantage of recent advances in digital software, utilising the extra ground gain available to an antenna when the moon is near the horizon, and carefully scheduling EME operations around optimum times of the month, successful contacts can be made by modest-sized stations.

Operating EME is relatively straightforward using the same type of equipment and software as other VHF and UHF bands. The ready availability of transceivers and amplifiers that operate on 6m, along with improvements in antenna designs and coaxial cables, bring 6m EME capability within the grasp of anyone interested in trying it, albeit it will require rather more effort than working the other propagation methods. So, rather than complaining when there is no ionospheric propagation on 6m, why not aim your antenna at the moon and see what is possible?

The following paragraphs cover the basics. For further information, there is plenty of material on the Internet. W7GJ's **[12]** website is a great starting point. For those with an archive of *QST* material or members' access to the ARRL website, the June 2005 issue also carried a primer on using JT65 for EME **[13]**.

While it is quite possible to make EME contacts on 432MHz on CW using a relatively modest station, CW is only feasible on 6m with exceedingly large EME setups at both ends of the contact (a few such stations do exist). What WSJT has done is to bring 6m EME to the point where it is no longer the preserve of those with lots of real estate and / or deep pockets.

The Logistics

The mean distance between the earth and the moon is 385,000km. The moon's diameter is 3476km so a little trigonometry shows it appears a mere 0.52° wide as viewed from earth. Bearing in mind the three-dimensional polar diagrams of typical amateur antenna arrays, it is obvious that only a very small amount of ERP will illuminate the disk. Moreover, the moon is a sphere, not a flat mirror, so only radio waves that hit the middle region will be reflected back to earth.

In other words, the path loss is enormous. To put some figures on it, the minimum round-trip path loss at 144MHz is 251.5dB rising to 270.5dB at 1296MHz. On 6m the overall path loss will be around 242dB, somewhat less than on those higher bands, but still considerable. But in practice the figure is somewhat academic as there as so many other factors at work, with 6m being affected much more by ionospheric factors than the higher bands.

Getting Going on 6m EME

There are three areas that deserve consideration in order to ensure successful 6m EME contacts: equipment, planning and operation.

Failure to assemble a station that can effectively receive and transmit enough dB will result in lack of success. By the same token, if you can exceed the threshold required for success, you will suddenly be rewarded with the capability to make world-wide contacts on 6m at any time of year or during any point in the sunspot cycle.

The moonbounce antennas at G3WOS, showing how they can be elevated.

One of the big challenges with 6m EME is the high level of urban noise if you live in a built-up area. This can make EME QSOs difficult with lower gain antennas, so living in the countryside is a very definite benefit! Of course, the main way of minimising this problem is to add elevation to your antennas, though this is no mean challenge with the large high-gain antennas required to undertake 6m EME. Equipment considerations have been covered earlier in the chapter as they are, to a large extent, common to all WSJT operations (albeit that EME requires as much power as you can muster). But the antenna system requirements for EME are somewhat more demanding than for any other 6m activity.

First you must decide what type of antenna you will use. It is essential that a modern, computer-optimised design of Yagi is selected. There can be many dB difference between specific antennas, and on EME (especially on 6m EME), every single dB is critical. It is a waste of time to try to use one or two old Yagis just because they happen to be available behind the garage. Such a convenient antenna may provide acceptable results on strong Es or F2 propagation, but EME is a much more critical and far less forgiving judge!

The typical single Yagi successfully used on 6m EME for horizon-only contacts is a long boom computer-optimised 7-element Yagi. One of the main UK antenna manufacturers, InnovAntennas **[14]**, has many highly-optimised 6m antennas of all sizes to suit 6m EME needs. Other excellent antennas include the PowAbeam models designed by Derek Hilleard, G4CQM, and sold by The DX Shop. And there are excellent models available around the world too, for example from M2 in the USA. There is also some great information on the G3WOS website, where he discusses why and how he replaced his 6m antennas, mainly to optimise performance for EME **[15]**.

If you are going to limit your activity to an hour or two around moonrise and moonset, and your antenna is going to be mounted close to the ground (less than 25m high) you probably will be much more successful by selecting the largest single Yagi possible. This will give you the lowest possible angle of radiation (essential for assuring 'common moon windows' with distant stations on the opposite side of the earth), and will provide the best utilisation of the much-needed 'ground gain' (up to 6dB) that is afforded to a Yagi aimed at the horizon. Ground gain is an essential element in the overall system performance of a modest EME station. When beaming at the horizon (moonrise and moonset) reflections from the ground in front of your antenna will add to the direct signal and give another 3 – 6dB or thereabouts, at certain elevations and depending on the local terrain, the ground conductivity and the height of the antenna above ground. Ground gain can make all the difference to achieving a solid contact.

So, the more modest stations don't try to track the moon but, instead, wait for the moon to move through the main lobe of their antenna. (Note that, unfortunately, ground gain is considerably reduced when you live in a typical UK urban environment with houses in all directions).

If you are planning to track the moon for more flexibility in scheduling other stations, you will need to assemble an array to make up for the lack of ground gain afforded by the single Yagi aimed at the horizon. You will also need some way to point the antenna upwards and to read the elevation. In the simplest arrangement, this is provided by ropes and a protractor with a weighted string mounted to its centre. In general terms, it requires an array of four Yagis when elevated to match the gain of a single Yagi aimed on the horizon. This may be strictly correct from an antenna gain perspective, but elevation can reduce local noise by 8 to 12dB, vastly increasing the ability to decode JT65A signals. This is especially the case in typical UK urban environments.

Regardless of the type of antenna(s) you are using, it is essential that you use the lowest loss coaxial feedline possible, and that it be as short as possible. A good goal is to have the loss at 50MHz be under 1dB. Such goals are usually quite achievable with the new coaxial cables that are currently available (such as 4-50 Heliax) or by using 'hardline' for most of the run. If you are planning to use the cable in an application which will require repeated flexing (such as around a rotator), make sure you use one with stranded centre conductor (such as the LMR Ultra Flex) that is designed for such an application. The feedline data in Chapter 3 should help in selecting a suitable feedline for your location.

Planning

Because of the numerous variables involving different radios, computers and interfaces for them, feedlines and interconnecting cables, the complete station should be tested thoroughly before it is transferred to its final location or any attempt made at 'real' QSOs.

The specific days of the month that you choose to be operational can have a very significant bearing on your success. The EME signal 'degradation factor' changes continually as the moon progresses in its orbit around the earth each month. This factor varies depending on the distance to the moon and the amount of 50MHz radio noise being generated in the region of space behind the moon. The degradation factor is expressed in terms of additional dB loss, compared with the ideal conditions of the moon being at perigee (its closest point to earth), with a quiet background sky.

If you are constrained to a certain day because you are arranging a schedule with another station who also has a horizon-only antenna system, you will be at the mercy of the degradation on that particular day. However, if you are planning a DXpedition or have a steerable antenna, you have the ability to choose your operation times so they are best suited to days with the least degradation.

GM4JJJ's MoonSked software **[16]** is invaluable in planning the best times for EME communication. The Moon Graph output (see **Fig 6.3**) plots key factors including earth-moon distance (the smaller the better – the distance varies over time as the moon's orbit is not an exact circle), TSky (the noise temperature of the 'sky', which varies throughout the year and will affect signal absorption), and degradation (already mentioned, a calculation taking various effects into account). In an ideal scenario, all would be optimum on the same date or dates – in practice this rarely happens!

6m EME signals can also be affected easily by geomagnetic disturbances. Even if the MUF is not high enough to produce ionospheric communication paths at 50MHz, ionisation can cause the signals to be deflected off their course. This can happen as they are heading out through the ionosphere on the way to the moon and / or as the incoming signals are returning to earth. Many times, it is not possible to predict when such ionisation might cause QSB or even complete signal blockage. However, in some cases (because the sun rotates every 27 days), it is possible to predict potential auroral disturbances that can affect high-latitude stations.

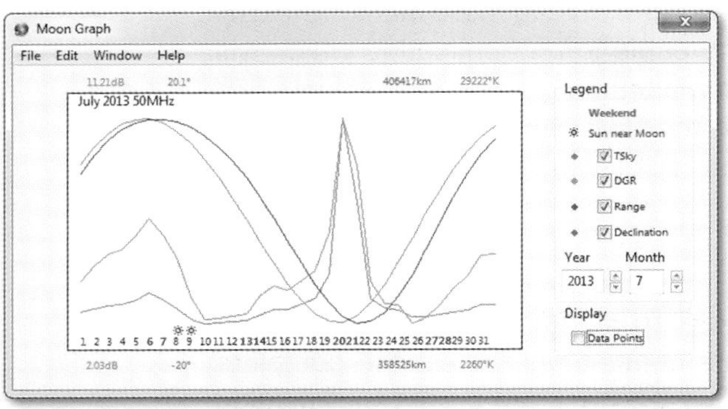

Fig 6.3: Moon Graph, from GM4JJJ's MoonSked software, shows the earth-moon distance, the sky noise temperature, degradation etc, plotted for each day of the month.

Similarly, during the spring and autumn, it is possible to predict the positions of high ionisation that can cause TEP propagation and can upset EME propagation if one or both stations are located in the lower latitudes (the 'TEP zone'). Intense Es openings can also block the EME signals, and the typical azimuthal directions of such ionospheric aberrations often can be approximately predicted depending on the time of day and time of year.

Certainly, if the local noise is high at certain times of day, that can be a serious consideration for a receiving station. For example, operating in the evenings, when all your neighbours' TVs and computers are turned on, may be problematic and early morning skeds are often the best for minimum noise and least chance of TVI.

Another example of a local factor would be the 'trapping' of transmitted signals by tropospheric ducting or a local temperature inversion. The effects of such atmospheric effects are most noticeable on signals when they are transmitted at low angles along the horizon, and are minimised if the signals can be transmitted at higher angles.

Successful 6m EME operations are often characterised by those that have the ability to provide real-time or near real-time coordination. When both stations have access to the Internet, it is easy to decide on a clear frequency and confirm that both stations are operational and can be available to start a schedule but this is often more difficult with DXpeditions which often depend on e-mail or satellite phones to pass on information regarding intended frequencies, transmitting sequences, and dates / times they will be operational. Because EME stations at one end of the path will often find that their chance to contact the other station is during the middle of the night, it is less compelling to gear up to try for a contact without any assurance that the other station will in fact be operational on a certain frequency.

Operating Considerations

Especially on DXpeditions, special RF filters on audio leads as well as scheduling and frequency coordination with other operators may be necessary to avoid QRM from other nearby transmitters.

The location of the antenna should be as close as possible to the operating position to reduce feedline loss. If the antenna is to be a horizon-only operation, special consideration must also be given to locating the antenna where it will have the most ground gain. There should be clear, unobstructed terrain in the direction of moonrise and / or moonset to maximise clean reflections coming up from the ground toward the antenna. If a clear view over a lake or the ocean can possibly be obtained, it is definitely worth the effort to locate the antenna to take advantage of that.

The standard mode for 6m EME operation is the WSJT mode of JT65A. It is very helpful to practice with the software and become experienced making contacts before the first EME operation. Practice sessions can be arranged with other local VHF stations or on HF. HF practice partners can be found at the JT4 / JT65 terrestrial chat room **[17]**.

It is recommended that you use the established EME procedures and standard messages, so as not to confuse your QSO partners or make it more difficult for them to copy you.

W7GJ adds this advice: "In the way that the standard messages are structured in WSJT, the contact is completed by sending the standard message following the one that you receive. For example, if you receive the

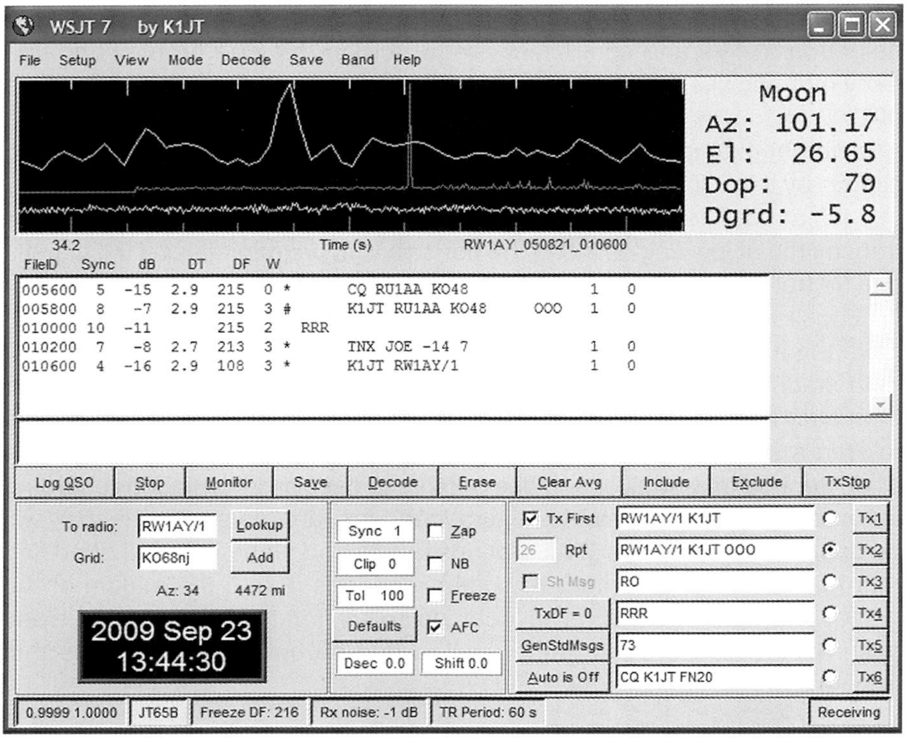

An EME QSO between K1JT and RU1AA using JT65.

first message (callsigns), you select the second message (callsigns and OOO reports) to be sent. If you receive the second standard message of callsigns and reports, you send the third message (RO), which indicates receipt of the callsigns (by sending the O report) and receipt of the report (by sending R). Especially if you have a limited moon window (caused by one or both stations being restricted to horizon-only operation), you cannot afford to make the mistake of transmitting back the same message that you have just received; the contact progresses by alternating messages between the two stations.

"It also is essential that both stations transmit during each of their scheduled sequences throughout the entire schedule period. Unlike ionospheric propagation, where it is often best practice to wait until you can copy the other station before you call them, on EME it is very unusual for both stations to be able to copy each other at the same time. This is because Faraday rotation changes the polarity of the signals going out through the earth's magnetic field to the moon and back. Therefore, one station may be copying fine while signals are cross polarised at the other station, attenuating them by more than 20dB! That is the main reason schedules are run on EME – so that the required information can be sequentially exchanged between stations to eventually result in a contact."

Faraday Rotation and One-Way Propagation

This is not the place to go into huge technical detail about EME operation, as most readers of this book will probably never attempt it, and those who do will find help and resource material on the web and from dedicated 6m EME players. But it is worth being aware, in simple terms, that there are a lot of factors which can affect a signal on the long journey to the moon and back, some of which have already been mentioned.

In addition, the polarisation of the signal can be affected such that instances of apparent 'one-way propagation' occur, where the station at one end is able to decode the far end's signal OK, but the reverse is not the case.

On the EME path, polarisation can change quickly, in a matter of minutes, which can disrupt what was on course to be a perfectly good contact. The subject is covered on, for example, the W7GJ **[10]** and G3WOS **[16]** websites, and the most serious of 6m EME operators have installed antenna systems where they can switch polarisations as required or, since the introduction of MAP65 as part of the WSJT software suite, take signals from two orthogonally-polarised antennas and combine them using MAP65 to overcome Faraday rotation effects.

But it's a big challenge. It's bad enough having to put up a two- or four-Yagi array for EME, without having to double it up in terms of polarisation (although there are certainly some X-Pol dual-polarisation Yagis available nowadays, which make the challenge a little easier). As a result, although MAP65 is now being used regularly for 2m EME, at the time of writing no-one that I am aware of is using it on 6m, but serious 6m DXers are nothing if not ambitious, so this may well change in time. For a fuller explanation of Faraday rotation refer, for example, to Wikipedia where it is covered in some detail **[17]**.

Results

Don't expect to work the world on 6m EME, at least not immediately! However, there are a few very well-equipped four-antenna 6m EME stations who are active regularly and prepared to make skeds. Most of these folk have designed their stations with the aim of being able to work a 400-watt station with single Yagi at the far end.

Occasionally expeditions to rare locations will make an effort on 6m EME to make a particular DXCC entity available on 6m when terrestrial propagation is unlikely. And W7GJ works hard to put on EME DXpeditions to rare spots such as S79GJ (2019) and CP1GJ (also in 2019). Of course, it is unreasonable to expect a DXpedition operation to run anything more than a single Yagi, so if you aspire to work such expeditions you will need to aim for something like a multi-Yagi antenna at your end, along with as much power as your licence allows (some countries, including the UK, make higher-power permits available, for example, under certain circumstances and on application).

QSL from W7GJ's 2019 6m EME DXpedition to Bolivia.

Finally, check out Wikipedia **[18]** for more about EME, and take a look at the websites of IW5DHN **[19],** N1BUG **[20]** and DF9CY **[21]**, all of which can be recommended for those contemplating EME operation.

OTHER DATA MODES

In closing this chapter, it is worth a short review of a couple of other data modes that are often encountered on 6m. All interconnect with the radio and computer in the same way as described earlier in this chapter in the context of the WSJT software and, in the same way, are 100% duty-cycle modes during transmit, so beware that you adhere to the advice of your transceiver manufacturer in terms of how much power you can safely run. While some are not, strictly speaking, weak signal modes, they have much in common in that the PC is doing much of the work.

PSK & BPSK

While by no means a weak-signal mode in the same way as the various WSJT modes, PSK (and BPSK) is easier to use for 'ragchew-style' contacts and yet can be very effective at low power levels or when signals are weak (but, generally speaking they do need to be audible; don't expect them to recover signals that are well below the noise level in the same way as WSJT). PSK is used quite extensively on both 6m and 4m as a search for data modes spots on DX Summit, for example, will reveal.

BPSK stands for Binary Phase Shift Keying and the most popular amateur system is BPSK31 (usually referred to simply as PSK31) **[22]**. It was developed by Peter Martinez, G3PLX, specifically as a replacement for traditional RTTY for real-time contacts. Peter's argument was that modes using error correction were inherently unsuited for real-time contacts, not for technical reasons, but because the users tended to be uncomfortable with the time delays inherent in hand-shaking between the two ends of the contact. The '31' in the name refers to the number of characters sent per second. The character set is based on a number of 'varicodes', i.e. variable-length codes, devised in a similar way to Morse code, whereby the more common letters can be sent more quickly than the less common ones.

Without going into any more detail, the benefits to the user of PSK31 compared with traditional RTTY are that it tends to work better in weak signal conditions and uses less bandwidth (a single carrier, as against the two of RTTY, with a theoretical bandwidth of 31Hz).

PSK31 requires a stable transmitter and receiver, but this is not a problem with modern transceivers. In any case, most of the software available for PSK31 uses a waterfall display and incorporates DSP filtering. Therefore operating the mode is simple. You can leave your transceiver's filters on a wide (SSB) setting and watch something like 2.5kHz of the band. This is room aplenty for many PSK31 signals. You can then click on an individual signal to decode it and start a QSO. Unlike RTTY, it doesn't matter whether you transmit and receive in USB or LSB mode, as there is only one tone, so there are no issues about tones being reversed. And if you invoke automatic frequency control (AFC) it will stay locked to the signal, even if there is some subsequent drifting in frequency.

PSK63 **[23]** is a variant of PSK31 operating, as the name suggests, at a higher speed. It has been developed very much with contesting in mind, with the intention that contacts can be conducted that much more quickly. As you would expect, it takes up roughly twice the bandwidth of PSK31.

Most of the popular software programs available for PSK31 also support PSK63. Programs include Digipan, WinPSK, Winwarbler, Multipsk, PSK31 Deluxe, QuickPSK and RCKRTTY. As well as text, PSK63 is capable of sending thumbnail pictures, with a transmission time around two minutes.

JS8Call

JS8Call by KN4CRD **[24]** is an experimental data mode based upon the underlying principles of FT8 but with a messaging and network protocol layer on top, to enable keyboard-to-keyboard QSOs to take place. It is intended primarily for HF use but lends itself to 6m and 4m 'ragchew' style QSOs.

Other Modes

While there are many other datamodes such as FLDIGI and FSQCall (and new ones appearing all the time), the only other mode encountered with any regularity on 6m and 4m (although not so much so nowadays) is traditional RTTY. While something of an anachronism nowadays, RTTY enjoys huge popularity on the HF bands as it still lends itself better than most data modes to quick-fire two-way exchanges. Perhaps it is because there are so many operators familiar with RTTY that it survives on 6m and 4m in the face of more recent and sophisticated data mode programs.

REFERENCES

[1] WSJT page: http://physics.princeton.edu/pulsar/K1JT

[2] ZL2IFB FT8 Guide: www.g4ifb.com/FT8_Hinson_tips_for_HF_DXers.pdf

[3] *QST* Articles explaining WSJT: 'Work the World with WSJT-X, Part 1: Operating Capabilities' (a general introduction to the digital modes supported by WSJT-X), *QST*, October 2017, pp30 – 36. 'Work the World with WSJT-X, Part 2: Codes, Modes, and Cooperative Software Development' (dips further into the mathematics, Forward Error Correction etc), *QST*, November 2017, pp34 – 39.

[4] 'The FT4 and FT8 Communication Protocols', Steve Franke, K9AN; Bill Sommerville, G4WJS and Joe Taylor, K1JT, *QEX*, July / August 2020: https://physics.princeton.edu/pulsar/k1jt/FT4_FT8_QEX.pdf

[5] 'The MSK144 Protocol for Meteor-Scatter Communication', Steve Franke, K9AN and Joe Taylor, K1JT, *QEX*, July / August 2017: https://physics.princeton.edu/pulsar/k1jt/MSK144_Protocol_QEX.pdf

[6] Dimension 4: www.thinkman.com/dimension4/download.htm

[7] ON4KST: www.on4kst.com/chat/start.php

[8] Ping Jockey: www.pingjockey.net/cgi-bin/pingtalk

[9] International Meteor Organisation: www.imo.net

[10] DL1DBC: www.dl1dbc.net/Meteorscatter

[11] Wikipedia on EME: http://en.wikipedia.org/wiki/EME_(communications)

[12] W7GJ: www.bigskyspaces.com/w7gj

[13] 'EME with JT65', Gene Zimmerman, W3ZZ, *QST*, June 2005, pp80 – 82.

[14] InnovAntennas: http://innovantennas.com

[15] G3WOS antennas: www.gare.co.uk/6m_antenna

[16] GM4JJJ MoonSked: www.gm4jjj.co.uk/MoonSked/moonsked.htm

[17] JT4 / JT65 Terrestrial Link (chat page): www.chris.org/cgi-bin/jt65talk

[18] Wikipedia on Faraday rotation: http://en.wikipedia.org/wiki/Faraday_rotation

[19] IW5DHN: www.qsl.net/iw5dhn/index.html

[20] N1BUG: www.g1ogy.com/www.n1bug.net/prop/eme.html

[21] DF9CY: www.df9cy.de/serendipity/index.php

[22] PSK31 'Official' Homepage: http://bipt106.bi.ehu.es/psk31.html

[23] PSK63: www.qsl.net/kh6ty/psk63

[24] JS8Call: http://js8call.com

Chapter 7

6m & 4m Around the World

IN THIS CHAPTER WE TAKE a look at some of the DX achievements around the world, including some of those who are close to the top of the 6m DXCC listings.

The 6m and 4m DXer has to have a number of psychological characteristics such as patience, persistence and a strong level of motivation. Exactly how these play out in practice depends on where in the world you live. If you live in one of the main centres of 6m activity (US, Japan, Western Europe) or 4m activity (principally Europe for the moment) you can, should you so wish, rely nowadays on Internet-based sources for DX alerts. The various chat rooms and Cluster systems which are discussed elsewhere in this book can provide the necessary alerts to much of the DX that is around.

But that doesn't mean that you will necessarily hear the DX or, if you do, that you will be able to work it. Propagation, as discussed elsewhere in these pages, can be very localised and your near neighbour may be able to hear a DX station that is completely inaudible with you. So even when you are alerted to an interesting DX station you may still have to sit on his frequency for quite some time before his signal emerges from the noise (or perhaps doesn't!) And if it does, because you live in an area of high 6m population, you may well be competing with lots of other stations to make a contact during what might turn out to be a very short opening. It can be quite a challenge.

In contrast, if you live somewhere like South America, Southern Africa or Oceania, the challenge can be very different. There probably won't be others in your area busy spotting the DX and you will have to hone your technique for finding it under your own steam. Typically this might mean programming your VHF receiver with a selection of commercial stations and beacons (VHF TV or radio carriers, for example) in the direction in which you are interested and scanning these continually so that you can immediately notice when the MUF starts to climb in that direction, or, as was mentioned in the chapter on equipment, using a modern SDR receiver to monitor key parts of the spectrum below 50MHz.

That said, in this age of FT8 activity, the playing field has changed somewhat. During the Es 'season' (very largely May through August), an active DXer will watch 50.313MHz, 50.323MHz or in the case of 4m, 70.100MHz, 70.154Mhz or 70.174MHz (varies by country), perhaps on an omnidirectional antenna, watching for openings. This may be all that is needed to alert him to possible DX. And the serious 6m or 4m operator will also use the other tools in his arsenal, including meteor scatter and, if he has the means to do so, moonbounce. While, of course, looking and hoping for a return of sufficient sunspots to give rise to F layer propagation – something that has been sadly lacking during Cycle 24. As for cycle 25, which officially started in December 2019, predictions vary widely but we can but hope.

HIGH ACHIEVERS

It's interesting to look at the DXCC standings for 6m, **Table 7.1**, **[1]**. The first nine are from Europe, perhaps not surprising given that European amateurs have a head start with something like 70 DXCC entities on their doorstep. Place number 10 goes to JA1BK and place number 15 to an African station, EA8AK. PY5CC is the highest South American, in 22nd place, while the highest placed North American on the list is K2ZD at 55th position (and this despite the fact the USA has had the 6m band for much longer than most European countries). So, it's undoubtedly more of a challenge to gain high DXCC standings from North America. Indeed, in Montana, where W7GJ is located, even Sporadic E is rare which is why Lance has become focused on EME operation (thankfully, there are wide open spaces in western USA where amateurs can put up some big antenna systems).

Of course, it is even more challenging from VK, ZL and the Pacific. So, we can't necessarily use standings in the DXCC program to determine who is the most 'successful' 6m DXer because geography clearly plays a big

Posn	Callsign	DXCC confirmed	Posn	Callsign	DXCC confirmed
1	LZ2CC	280	14	IK4DRY	243
2	ON4IQ	268	15	EA8AK	242
3	ON4GG	265	16 =	9A8A	241
4	G8BCG	257	16 =	CT1EEB	241
5 =	EA7X	254	18	SP4MPB	237
5 =	IK0FTA	254	19	SM7FJE	234
5 =	SV1DH	254	20 =	DL7AV	233
8	DL8YHR	252	20 =	IZ4BEZ	233
9	9H1BT	249	22 =	G0JHC	230
10	JA1BK	248	22 =	PY5CC	230
11	IW9HII	247	24	S51DI	228
12	G3WOS	245	25 =	G8VR	227
13	LZ2DF	244	25 =	I4EAT	227

Table 7.1: The top 25 DXCC standings on 6m, as of 25 September 2020 (from ARRL DXCC – 6 meters).

part too. Nevertheless, I thought it would be interesting to ask some of the high achievers to share with readers how they got to where they are. I dropped a line to the top 10 on the DXCC standings and I'm delighted that several of them were kind enough to share their stories. So, here they are.

Geert van de Velde, ON4GG

(Position 3 with 265 credited)

Geert is currently using 4 x 9-element M2, stacking 1/2in hardline, 1-5/8in feedline, and says that every 0.1dB counts! He has been using this system for about 20 years now although he has used many radios over the years (currently a TS-890). He says he has been on the band since 1991 and spending "16 hours a day on the radio", working tropo, meteor scatter, aurora, long-path and with about 20 worked only on EME. He also credits a good hilltop location. Not surprisingly, he agrees that it is essential to be on the digital modes nowadays to work new ones. And Geert is also preparing a serious (4 x 15-element) system for the 4m band!

Peter Taylor, G8BCG

(Position 4 with 257 credited and 3 yet to be claimed)

Peter has held the UK callsign G8BCG since 1967and his main interest has always been VHF and UHF long distance propagation.

In January 1979 he moved from Manchester to Honiara, Solomon Islands. Within six weeks of setting up home in Honiara he was on 6m with 6 watts from a Kenwood TV-506 to a quad loop slung in a tree. Peter says, "This was VHF like I had never heard – I was hooked. A good solar cycle, excellent tropical location and a sea path to Japan combined to entrap me for life! 50,000 JA QSOs later and I was still looking for more when I left H44 in 1986."

Peter, G8BCG, operating as H44PT from the Solomon Islands during Solar Cycle 21.

The next 10 years or so in the telecommunications industry saw him travelling all over the world, including extended periods in Thailand, Australia, the Solomon Islands, California, South Korea and southern Africa. As he says, "Not so great for ham radio but I did manage to get on the air whenever possible – you guessed it – on 6m!"

After returning from W6 Peter moved to Cornwall and has been on 50MHz from IO70 since 1997, gaining his 50MHz DXCC in 2001.

His VHF station is housed in a secure site container unit on the top

G8BCG's EME antennas: 4 x 8-element InnovAntennas OP-DES Yagis on 6m.

of a small hill behind his house at 160m ASL and he is about 6km in from the coast between Looe and Polperro in Cornwall. Everything on site runs from solar plus a 25kVA diesel generator. 4G Cellular provides Internet and security webcams etc.

His current 6m setup is: IC-7800 + GU84B PA, 4 x 8-element OP-DES Yagis from InnovAntennas on a 6.5m x 6.0m H-frame with full Az / El on a 32m tower (usually kept around 20m).

Despite having LDF450 / Ecoflex 15 feeders, Peter also uses a mast head pre-amp – this makes a much larger difference than one would expect, he says, particularly on EME. The main advantage is that it increases the receive signal in the shack by 20dB, thus improving noise immunity for in-shack systems. He is fortunate to be in a low noise environment that has not yet been overrun with industrial solar PV or ADSL.

Peter's DXCC total of 260 breaks down as CW 177, SSB 191, data 196, terrestrial 226, EME 96 and a total of 23,429 6m QSOs in 38 zones, 132 fields and 1273 squares.

Peter says, "I am very fortunate to live in a noise-free rural environment and have space for antennas. The south-west coastal location is also good for access to the maritime ducting to the south and west and of course for TEP on 6m. The site is rather exposed, so my towers and large arrays spend a lot of time luffed over and tied down in the winter".

And as far as modes are concerned, "I love SSB and CW but the reality is that data modes have transformed our hobby. 6m activity has rocketed and there is always someone to be found on the band. Weak signal performance combined with multi-user single channel operation and automated activity reporting via the Internet has led directly to the 'discovery' of previously unnoticed propagation and paths.

"A good example of this is the southern hemisphere version of Summer Solstice multi-hop Es propagation between VK / ZL and South America –

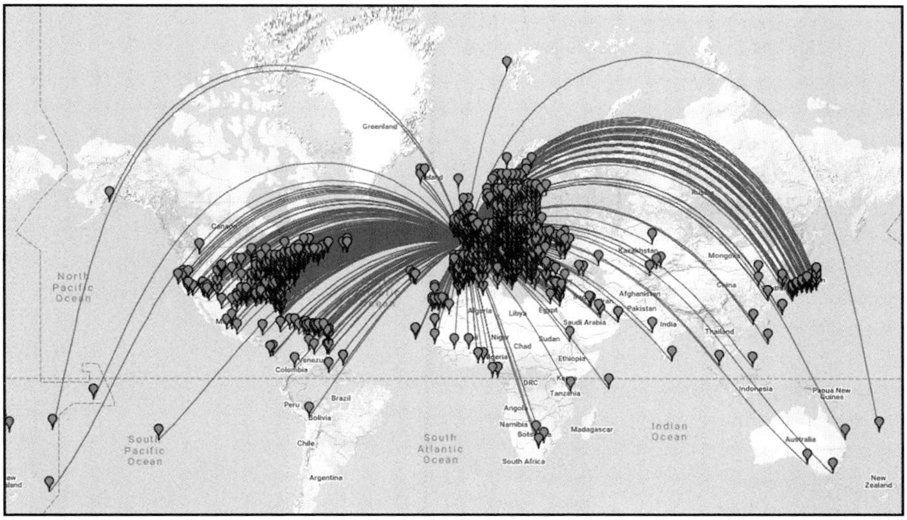

G8BCG's 2019 6m QSO map.

now a regular on the DX calendar. Of course, on 6m JT65 has brought EME to the masses! Single Yagi stations with relatively low power can work EME and DXpeditions are increasingly making use of this with a single long Yagi and SSPA (solid-state PA) to enable new 'firsts' from rare entities.

"For me, data modes have changed the hobby – but they are certainly not killing it – in fact quite the opposite. Again, for me, it's not about the mode, it's about identifying the propagation, then finding and working the DX whenever and however that is possible!"

And while Peter has full legal power capability and a 4 x 7 LFA array on 4m, the reality is that his passion for 6m precludes serious DXing on 4m!

Sergio Roca, IK0FTA

(Position 5= with 254 credited)

Sergio uses an IC-7800 and a 6-element Yagi from Create of Japan. He has been QRV on F2 from 1991 to 2002, during which time he worked about 200 countries, something that those who have recently arrived on the band would be envious of! It's also true that southern Europe probably enjoys (or enjoyed, at least) longer and more extensive F2 openings than those further north. Although Sergio has made a handful of EME QSOs, none of them were 'new' countries and he says local noise makes EME very difficult for him (he lives near the centre of Rome, about 2km from the National Palace and close to many offices, restaurants and shops).

Sergio is unhappy about the current prevalence of data modes but recognises that maybe 95% of DX activity is on those modes nowadays.

Right now Italy has no access to the 4m band but, in the past when temporary permits were available, he worked about 38 countries (including TEP to ZS6) and was also active from 1A0KM and HV0A.

Costas Fimerelis, SV1DH, in his shack in Athens.

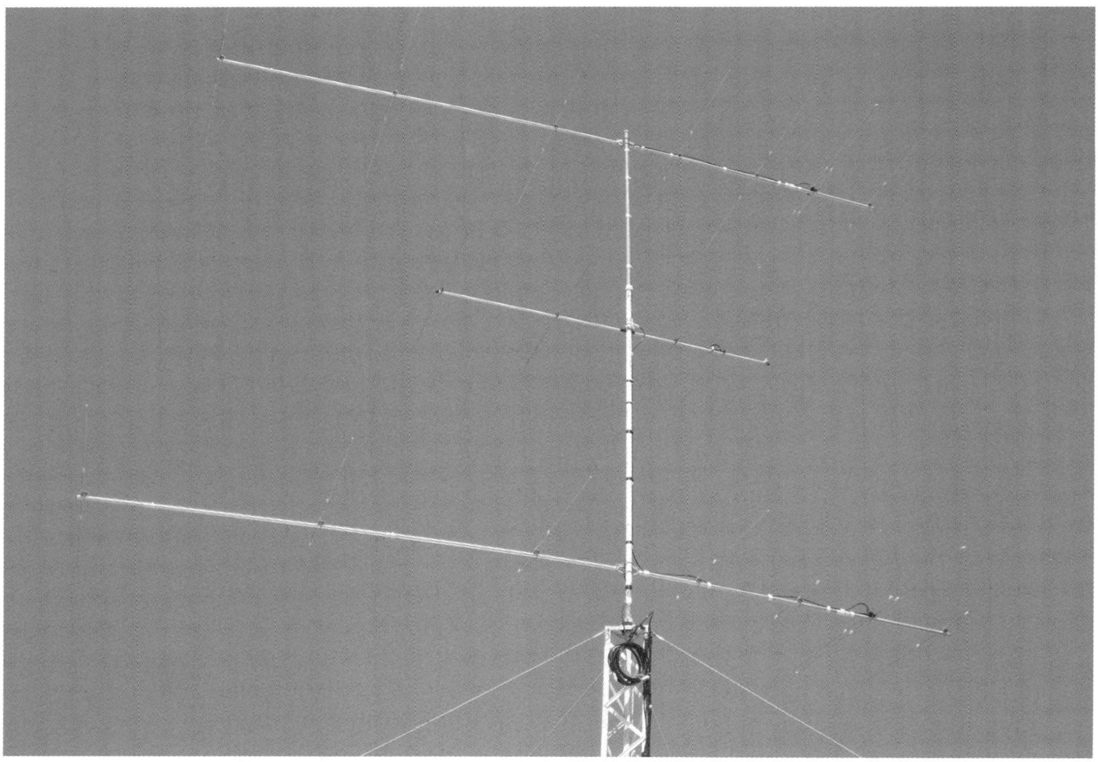

SV1DH's antennas: a stack of two 7-element 6m Yagis and a 5-element Yagi for 4m.

unit f 14 stadium way
bletchley milton keynes
MK1 1ST
tel: 01908 229248
vat reg: gb 340 377 862

15518 Vyshak

CHK 6909 TBL 831/1 GST 1
12 Oct'24 16:14

---

1 high five	0.00
high five large	6.00
1 sparkling water small	3.00
1 chicken katsu curry	14.00
3.83 vat 20%	23.00
Net Total:	£19.17
sub total	£23.00
total	**£23.00**

------service not included------

ank you

soul club. loyalty has landed
join the community today
100% of tips are shared across this
restaurants team
enjoyed your noodles? please leave us a
review

The SV1DH QTH.

Costas Fimerelis, SV1DH

Position 5= with 254 credited)

Costas has been QRV on Six since 1980, enjoying both F2 short-path and long-path propagation in the days of sunspots, along with Transequatorial Propagation (TEP), Es, ionoscatter, meteor scatter and tropo. He is currently using an Icom-7700 and a Kenwood TS-480, a single Create CL6DXX 7-element Yagi and a vertical stack of two of the same. Of his total country count, all but two were terrestrial.

Costas lives in Athens where he is surrounded by mountains and with high local noise. However, he has a remote summer QTH, which is much less noisy and with a clear horizon to easterly QTFs.

He values the use of data modes to work the weaker DX during Es openings.

He is also QRV on 4m, using an Icom IC-7300 and 5-element Yagi.

Frank Hobelmann, DL8YHR

(Position 7 with 252 credited)

Frank has been on the band for 30 years. He uses an IC-7800 and, in the past, 4 x 5-elements from M-squared. However, he has recently changed this to a horizontal stack of two 8-element Yagis from Antenna-Amplifiers. He says this is because most activity is now on data modes (which, nevertheless, he considers a pity) but his new system is optimised for 50.313MHz.

DL8YHR's antennas: a horizontal stack of two 8-element Yagis from Antenna-Amplifiers.

As well as his station, Frank credits his high score to the fact that he works in his own garage, just two minutes from his radio room.

Frank is looking forward to (hoping for) some of those great F2 conditions that we have seen in the past, with openings to KH6, V7 and similar.

Meanwhile, Frank has recently started on 4m, with an IC-7300 and 7-element Yagi.

Paul Galea, 9H1BT.

Paul Galea, 9H1BT

(Position 9 with 249 credited)

Like many old timers (licensed: October 1969) most of Paul's equipment is home made. He uses a Kenwood TS-940 as the 28MHz IF for all the 50, 70 and 144MHz transverters. In 1974, when still a student, he managed to re-engineer a transverter for 144, based on the commercial Europa B transverter with a 28 – 144 mixer (6360), driver (6360) and final (QQV06-40A). This was driven by a TS-515 on 28MHz and the transverter gave about 100 watts output. Generally, all subsequent transverters were similar, just changing the LO, the receive converter and the interstage coils according to the band in use. He is still using these transverters mainly because he has a good supply of 6360s and two spare QQV06-40A in the junk box.

When conditions deteriorated, and most of EU was active on 6m, new DX was hard to find and became difficult to work so he had to resort to some more power giving him a little bit more punch on the pile-ups. He was also able to try EME on moon rise / set with no elevation.

Paul started with a 4-element Yagi then tried a 6-element NSB design and finally for the last 15 years used a home-made copy of the 6M7JHV 7-element Yagi and a 10m run of LFD4-50 coax. As the coax length was short he did not use any masthead preamp, mainly because he has always been plagued with neighbourhood QRN.

In Malta activity on 50MHz started on 6 June 1987 when the then Wireless and Telegraphy Department granted permission to local amateurs to use the 6m band. Alain Wright, GW3LDH, and his wife Maureen had, on frequent visits to Malta, encouraged the local amateur radio clubs to push forward and submit an official request to allow activity on this band. Paul started on the band the day permission was granted.

This, he says, started a very long, exciting and rewarding experience on the 'Magic Band'. "We immediately started having extensive Es openings to G, GM, GW, EI. On 17 June 1987 we had a brief trans-Atlantic opening with the usual big guns K1LPS, K1TOL, etc. In October '87 FY7THF/B came in and subsequently a few days later I worked FY4EE. I also started having TEP QSOs with ZS4TX, A22KZ and ZS3E.

"All propagation died down during the winter months. In the meantime Alain, GW3LDH / 9H1AW, brought over a 50MHz beacon, which was sited at the MARL club. On 3 March 1988 I received an unexpected phone call at about 2045UTC from Mike, ZD8MB, who was receiving the Malta beacon. Within five minutes we were in contact on 50.110MHz. We eventually established a daily sked at about 2030UTC which for the next couple of months never failed. In fact, we found out that propagation always started around the same time: from S0 to S9 in 15 minutes at around 2030UTC and then dwindled slow back to S0 by 2330UTC.

"One evening during the sked I received a call from LU8MBL on the same beam heading. I kept the daily sked with Mike throughout March and April, sometimes joined in by South American stations, mainly LUs and PYs.

"As the TEP improved towards South Africa, the south-west path to ZD8 / LU / PY was lost and again by the end of May as Es picked up we lost the TEP to South Africa. This pattern was repeated in subsequent years.

"The summer Es period gave us hundreds of QSOs with G stations. In June we had a few openings to the US. In October 1988 we started again with the usual afternoon TEP opening to South Africa and in the late evening the TEP path opening to South America. On 8 October 1988 during a very good opening to South America, at 2145UTC on the same south-west beam heading, I worked JA2DDN 54 / 54 followed by a huge pile-up of very well-disciplined JA stations. By 2330UTC I had over 60 JA stations in my logbook.

"On the 9th we again started at noon with the TEP to South Africa followed by a brief opening to South America at 2100UTC and at 2130UTC another

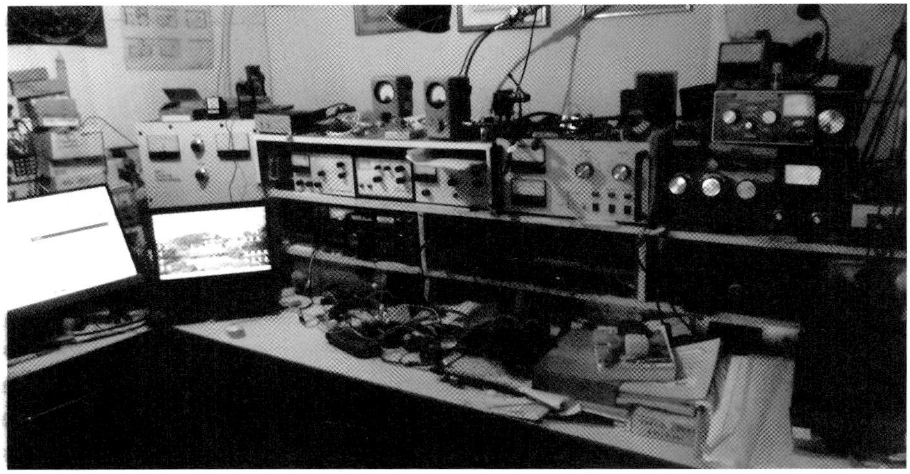

The well-equipped workshop and shack of Paul Galea, 9H1BT.

opening to JA with more than 80 QSOs in the log. During the period October / November we had the afternoon TEP to the south and the late evening TEP to the south-west, but no more JA openings.

"In late November 1988 and January '89 we had Es openings to USA and VE. During February we had the odd brief opening to different areas working ZS, TU, ZD7, HC5, OA8, J52, CX4, TR8, FY, KP2, VE1, HH7, FY5, 8R1, LU, PY. On 26 February I had a good opening to VS6 [Hong Kong, now VR2 – *Ed*] and at 1115UTC logged the OX3VHF beacon. On 5 March there was a brief opening at about 1200UTC to FY and HC. In hindsight I think these were F2 openings.

"On 7 March was the usual late evening opening to the south-west, and at 2338UTC another opening to JA up to 0127UTC on the 8th. I could not take it any longer so went to bed, but woke up at about 0400UTC and could still hear JA stations on the long path. On the 24th we had a similar pattern of openings starting with TEP to the south then south-west then a long path opening to JA followed.

"On 25 March 1989 at 0933UTC I worked VK4VV completing my WAC on 6m. This was followed by a few VS6 stations and a few JAs via short path. In April we had a few brief openings via TEP to the south-west and in May the usual Es openings to EU. In summer I had many Es QSOs on 2m, however, on 7 July we had a very good opening to US / Canada working about 50 stations in that area. In August / September was the late afternoon TEP to South Africa and the late opening to South America .

"On 14 October at 0800 – 0930UTC very strong signals to KH6 with the beam directly south; I must have worked all the stations active on 6m in Hawaii. On a side note on 18 October 1989 during a very good TEP opening to South Africa I managed to work ZS3E [Namibia, now V51E – *Ed*] on 144MHz on CW with 519 / 518 reports and later I heard Kosie working into Italy on SSB at 53.

The 9H1BT 6m and 4m antennas.

"The above is a pattern of events during subsequent solar cycles. We were very lucky to get permission to operate on 50MHz just at the beginning of solar cycle 22. However, during 1994 – 1997 (the cycle minimum), I only managed QSOs via Es, working a few new ones as more countries became active on 6m.

"TEP activity for Solar cycle 23, which started around September 1998, followed more or less a similar pattern as the previous cycle; afternoon / evening openings to the south-east, south and south-west to the Indian Ocean, South Africa and South America.

"It was only in 2000 that we again started to have the spectacular openings to Japan via long path at night with three major openings on 19, 20 and 22 March 2000 to JA. In October, on the 3rd, 4th and 6th again openings to JA via long path. During this cycle we had many openings to the Far East, Asia, Oceania and the west, mainly US and the Caribbean. In cycle 24, I have not recorded any late night openings via long path to Japan. The other TEP, F2, Es and multi-hop Es openings were similar to the previous cycles.

"All QSOs on 6m are terrestrial except two via EME, namely ZL and FK8. It took me three years of trials to work the FK8 but we finally made it during a late night and relatively quiet moon rise."

Paul lives in Dingli, the highest village on the western part the Maltese Islands at about 280m ASL. He has a very good take-off in almost all directions. To the west, south-west and south he is about 1km from the Dingli Cliffs with a sheer drop to sea level. In the other directions he has to go over the island, but as Malta is a rather small and relatively flat island, he doesn't notice any major degradation in the propagation on HF and the lower VHF bands. However, one problem in the last few years is high rise buildings in the neighbourhood and also that local noise can sometimes be very bad.

Paul notes that the days are gone when he could park on 50.110MHz and wait for the DX signals to pop up. It seems that everybody is on 50.313MHz. He believes there is a need to identify data frequencies for DX operation, away from 50.313MHz.

Paul has also been active on 70MHz since permission was granted on 16 February 2015. He built a transverter, put up a 6-element LFA Yagi and worked 43 counties with many firsts, including TEP QSOs with South Africa. However, although he still has a few countries to work given that activity is restricted only to Region 1 and not many countries are active his interest in this band has slowly dwindled.

SIX FROM JAPAN

(with thanks particularly to Han Higasa, JE1BMJ)

Japanese amateurs were granted the use of 6m in 1952, although there is evidence of activity in the 1937 – 38 period on 56MHz by stations in the J2 (Tokyo – Yokohama) area. It is possible that after WWII the first 6m QSOs were made by occupying Allied forces personnel, as Japanese nationals were prohibited from amateur radio activity until the end of the occupation in 1952.

Six is extremely popular in Japan and, for many amateurs, is their entry band to the hobby. There are some thousands of JA amateurs active solely on Six, and undoubtedly the advent of FT8 has increased the band's

One of Japan's top 6m DXers, Kazu Ogasawara, JA1RJU, seen here operating on his favourite band at 3B9C, Rodrigues Island.

popularity for DX chasing, not only because FT8 allows the smaller stations to work DX but because FT8 gets over any barriers of language.

Unlike Europe, there are not too many DXCC entities within one hop of Japan. Those that are include HL, DU, JD1 (Ogasawara), JD1(Minami Torishima), BV, BY, UA0, JT, XX9, VR2, KH2 and KH0.

TEP openings are usable from JA7 in the south of Japan, but rarely from JA8 in the north.

Skewed paths are almost certainly the result of scattering from high electron density regions around the equator. Okinawa, JR6, being south of the main Japanese islands, generally experiences better propagation than the rest of Japan, with more frequent openings.

JI1CQA's website **[2]** has a list of 6m 'firsts' from Japan and shows that 271 entities had been worked as of August 2018. A detailed history of 6m in Japan appears in English on JA1RJU's website **[3]**, translated from Japanese by Roy Waite, W9PQN.

SIX FROM NORTH AMERICA

(with thanks to the late Gene Zimmerman, W3ZZ, and updated by Chris Patterson, W3CMP)

The USA has had a 6m allocation from 50 to 54MHz since amateur radio resumed after WWII in 1945. Six metres is a popular operating spot for thousands of US amateurs and is aptly called the 'Magic Band' because of its eclectic propagation. The advent of HF / 6m transceivers has increased the level of activity. In recent VHF contests upwards of 4000 or more operators have participated on 6m. A significant number of these have stations and antennas capable of working long distances on 6m under marginal conditions.

The power limit on 6m is the same as it is on most other bands: 1500 watts output with no licence restrictions on antennas. Towers are generally limited to 61m height by aviation law without special permission and conditions. However, local zoning laws and Covenants, Conditions and Restrictions – 'CC&R' – may limit erection of towers and antennas. US amateurs use all forms of transmissions both analogue (SSB and CW) and digital.

A small number of stations have legal limit power and huge antenna systems with up to four 15m boom Yagis either stacked vertically or arrayed on an H-frame. Such large arrays include the multi-Yagi systems at K1TOL (in FN44), K2MUB (FN21), K3TKJ (FM28), contest arrays at K1WHS (FN43) and K8GP (FM18). These arrays can be used on EME and can work single Yagi stations with booms no more than 9m in length, particularly when both stations use JT65A.

The US is geographically large and diverse. Only a relatively small number of countries in the Caribbean, Central America and off the Canadian East Coast are reachable within the 2000km E-skip (Sporadic E) or the 5000km F2 single-hop radius. But, because of its size, essentially all types of VHF propagation can be seen at one time or another in many places throughout the country. Long distance contacts are possible via 'E-skip' propagation around the solstices and via F2 when the sunspot numbers reach levels high enough to generate solar flux indices over 200 for a few days at a time.

The more recently discovered propagation mode, SSSP (Summer Solstice Short Path), has enabled contacts between Japan, China and the eastern US in late June and July; the corresponding Southern Hemisphere mode has allowed contacts between Australia, New Zealand and other South Pacific entities and the US.

Transequatorial propagation at about the time of the equinoxes is possible via spread-F to South America directly for locations at about 30°N or below and via Es links to TEP for areas north of that.

Auroral propagation is common during periods of high geomagnetic activity in the more northerly latitudes especially above 40°N but can extend southwards to 30°N at K indices of 8 or 9.

Meteor scatter propagation over distances of 2000km is readily achievable everywhere and even tropospheric ducting is known, though it is much less frequent than on frequencies of 144MHz and above.

Propagation on 6m resembles that on 10m but is much more erratic because the MUF is twice as high. At the peak of the sunspot cycle much, but not all, of the world is workable from almost all of the US via F2. Within the US F2 propagation is relatively long and tends to bridge the east and west coasts both laterally and diagonally. During the summer months much of the northern hemisphere is available to some part of the US.

In general, Sporadic E is more common the farther south you go. Almost the entire contiguous US is reachable by two Es hops. During the equinoxes, South America is readily workable by TEP or Es-TEP links.

Some parts of the world, notably in central and south-east Asia, are very difficult to work from any part of the US because of propagation and perhaps because of lack of activity or licensing issues. Specifically, among these are Zones 17 – 19 (Asiatic Russia); Zone 22 (VU and 4S7); Zone 23 (JT and UA0Y); Zone 24 (BY, BV, VR2 etc), and Zone 26 (south-east Asia).

The east coast, particularly W1 – 3 and northern W4s, have problems working Japan and other northern Far Eastern countries. It is just in recent years that SSSP has been utilised to make these contacts. The west coast, and particularly the Pacific Northwest, has problems working Europe because the path is directly through the auroral zone. The Midwest also has problems to Europe because of the auroral zone.

The Caribbean and northern South America are easy to work from the east coast and the south, more difficult from the Midwest and not easy from the west coast, especially via Es because all contacts are multiple hop. The same is true for Europe. However, the path to Europe from Alaska is extraordinarily difficult and has been bridged only a handful via any mode of propagation. Until the past few years the Pacific Northwest has not been able to work Europe via Es but the advent of FT8 has made the path much more accessible.

SIX FROM SOUTH AFRICA

(with thanks to Chris Burger, ZS6EZ)

South African amateurs have had the use of 6m since the late 1940s and were able to work into Europe during that period when several European countries, albeit briefly, had 6m access (as chronicled elsewhere in this book). The list of 'firsts' on ZS6EZ's website **[4]** shows G and VE in 1947, for example, while the first JA – ZS QSO was recorded in 1958. But from then on until the late 1980s, when many more countries started to gain access to the band, South African amateurs had a fairly quiet time on Six, with even their near neighbours such as 3DA0, 7P and A2 being absent.

Sporadic E is notable in ZS largely by its absence, so there isn't the annual Es excitement that Japanese, European and North American amateurs enjoy. 6m activity is also concentrated in a relatively small area (about 1500km across), requiring high levels of ionisation before contacts become possible. Couple these constraints with the low number of active stations, and it is easy to understand why Es is not an everyday occurrence.

The latitude of ZS means that TEP is quite reliable and has been studied quite extensively, albeit more at 144MHz than at 50MHz. But the huge distance from ZS to the major centres of amateur radio population has meant either waiting for F-layer propagation to come along or to struggle during the sunspot minimum by working near-in DX via meteor scatter and the more distant DX by EME.

Even square chasing is harder than in Europe or North America, not only because of the lack of Es, but because the squares are larger! This is

Chris Burger, ZS6EZ, with editor Steve Telenius-Lowe, PJ4DX, during Chris's visit to Bonaire in November 2015.

because the nature of the QTH locator system is such that squares occupy a constant unit of longitude, but that gets narrower in absolute terms the nearer one gets to the poles (the effect is lost on a Mercator map projection). However, starting in the mid-80s, there was considerable square-bashing by a number of amateurs, resulting in firsts from most of the neighbouring countries. Most of these expeditions conducted meteor scatter schedules using SSB, serving a limited audience of perhaps a dozen DXers.

SIX FROM BRAZIL

(with thanks to Peter Sprengel, PP5XX / PY5CC)

Brazil is an ideal location for 6m operating, in a very good position relative to the geomagnetic equator. Don, PY5ZBU (ex-G3MWM), is believed to be the first person in the world to have confirmed 100 entities on 6m but his cards went astray *en route* to ARRL and Don lost interest in the band after that. So the first from South America was PY5CC who was at 200 countries by October 2000 (he currently has 230 credited).

Propagation from Brazil is largely via TEP, but because of the country's location, it is one of the first to enjoy F2 propagation when there is sufficient solar activity.

Amateurs elsewhere in the world are also in debt to the PY 6m guys for activating the three island groups that count separately from Brazil, i.e. PY0S (St Peter and St Paul Rocks), PY0T (Trindade) and PY0F (Fernando de Noronha).

There are only a handful of Brazilian amateurs active on 6m, but enough to ensure that Brazil is very workable from both Europe and North America on a regular basis.

SIX FROM AUSTRALIA

(compiled with help from Steve Gregory, VK3OT / VK3SIX, and the UKSMG archives)

6m in Australia dates back to the immediate post-war period, though there was a short period when the band changed to 56 – 60MHz. During the 1958 International Geophysical Year, a 50 – 54MHz allocation was reinstated, but from 1964 to 1989 this was reduced to 52 – 54MHz to avoid conflict with TV channel 13.

Australia is a big country, so the experience of 6m varies from one call area to another. But VK amateurs enjoy all the propagation types that have been discussed elsewhere in this book, including aurora (but via the 'southern lights' in their case). Stations in VK6 and VK8 are much more likely to enjoy F2 QSOs with Europe compared with those in VK2, VK3 and VK5. In contrast, the further north and east the VK station is located, the better the chances of working the USA.

What VKs do enjoy, along with the ZLs (and making European amateurs green with envy) are openings into the Pacific, to all those little islands (though, in fairness, 6m activity from some of those places is few and far between). But it's tough from VK and during sunspot minima EME can be the only way to catch new ones. Steve, VK3OT / VK3SIX, is the highest-placed VK on the DXCC listings, with 137 countries confirmed, a total that a well-equipped European 6m operator could achieve in just two or three seasons. In Steve's own words, "It always amuses me to see European operators comparing their country tallies. It's easy when you have 100 countries within range. I wish they would stop asking us how many countries we have worked this year!"

SIX FROM NEW ZEALAND

(contributed by Rod Mackintosh, ZL3NW)

Although there is recorded activity from 1934 within New Zealand on the then 5-metre band (60MHz) it was not until after WWII that activity increased on 6 metres (50 – 54MHz). There was wide availability of surplus valves such as the QV04/7, 832a, 815, 829b and 807. The 807 was used for both RF and modulators.

Modes then were CW and AM. Surplus equipment was also in abundance, for example RF26 units fed into R1155A, AR88 and ARC-5 receivers. Antennas were simple dipoles up to a maximum of 4-element Yagis.

It is documented that ZL4GY created a world record by working the USA in 1949. The USA callsign is not given but no doubt the station was on the west coast. Shortly after this, contacts were made from New Zealand to stations around the Pacific Rim. In 1953 Graham, ZL3GS, and Bob, ZL3NE, worked Wynn, ZL3DX, who was operating as VR2CG from Fiji (now 3D2). Also that year Graham worked Doug, VK9DB, in Papua Territory. This was Doug's first ZL contact. Looking at the log of Fred, ZL1ADP, it is interesting to see the level of activity when he was first active in 1957; there were clearly a lot more active 6-metre stations than today. In 1958 Fred worked his first JA, thought to be the first JA – ZL contact, and also his first USA station, K6RNQ. Most areas of Australia were worked at this time too.

The introduction of television and more particularly the channel 1 frequency allocation of 44 – 51MHz (video 45.25, sound 50.75) in 1959 limited the use of the 6m band in areas where channel 1 was, and still is, in use. Initially 6m could be used in these areas outside TV broadcast times. Later television was extended to 24 hours per day, and the introduction of permits limited the number of operators who may operate in the bottom megahertz of the band. These permits have only been issued to amateurs located outside the channel 1 TV area, though it should be noted that 51 – 53MHz may be used throughout the country. For some years activity was around 52MHz, with the calling frequency on 52.050, but nowadays the bottom 200kHz is widely used internationally and hence almost all ZL activity is also there.

Currently there are about 15 operators within New Zealand who are active on 6m, spread from the far north to Dunedin in the south. There are large areas in the country with no activity due to channel 1 TV. However, New Zealand channel 1 TV is a great propagation indicator for DX stations!

Although the Pacific Rim was worked as far back as 1949 to 1958, it was not until 1980 that this distance was extended, when Bob, ZL3NE, worked VE1AUX at a distance of 15,555km. On 3 April 2001 at 2003UTC Tim, ZL3VTV/1, made SSB contact with Jose, EH7KW, at a distance of 19,921km. Signals peaked at a direction of 145 – 150° and the signal strength peaked at S3.

A very interesting contact was also made on the same day when Raj, VU2ZAP, was worked long path by Ray, ZL2KT, and Morrie, ZL2AAA, at 2002UTC. Bob, ZL4AAA, in the far north also copied Raj and worked him the next day at 1947UTC at a beam heading of 160°. There was no signal on the direct path of 290 – 300° and Raj confirmed Bob's signal peaked via the long path of 315 – 320°. Raj was last heard at 2015UTC and peaked at S9.

During the better sunspot cycle peak years it is not unusual to have propagation follow the sun around the Pacific Rim. First thing in the day would be to look for any indicators of long path to Europe at about 150 – 170°. After this and a little later in the morning turn the beam to about 90° and listen for the XE1KK beacon. This beacon is an excellent propagation indicator and is often heard before an opening to the Caribbean, California or an extended opening to Florida.

Early afternoon look for any JA beacons, although as often as not the appearance of JA stations is the best indication that the band is open. Late afternoon, keep an ear open for any Hawaiian beacons. It is interesting that although the direction to Hawaii is similar to California the propagation times are quite different, i.e. there are different propagation modes involved. During the mid-afternoon there are often openings to VK1, 2, 3, 4, 5 and 7. Late afternoon or early evening an opening to VK6 and 8 and if one was very lucky openings to Europe mid to late evenings at about 270°.

During the sunspot minima years there is still some propagation via Sporadic E during December and January and to a lesser degree November and February. Contacts may be made to all of Australia and any South Pacific Islands which are active. The VK channel 0 TV is a great early indicator. For example, the video on 46.240MHz is often received first, followed by the sound on 51.740MHz. This is then often followed by the VK2RHV and VK2RSY beacons. At the peak of the Sporadic E season I have heard the VK6RSX beacon at Dampier for hours on end at a distance of some 5717km. Unfortunately, there are currently no active 6m stations in Dampier, but it would be interesting to know how far beyond the west coast of Australia propagation extends. About this same time John, VK6JJ, worked Paul, A35RK, at a distance of 6971km.

Winter Sporadic E typically offers propagation during June and July and there are openings to the east coast of Australia, VK1, 2, 3 and 7 with the occasional opening to VK4 and 5.

Six-metre EME was something I had not considered until Kjell, SM7BAE, e-mailed me in 2000. He knew I had recently put up a 10-element Yagi on a 13.2m boom at the top of my 25m tower and he considered a CW EME

The VHF antennas of Rod, ZL3NW. Right: 50MHz 10-element Yagi on a 13.2m-long boom.

contact may be possible. Back in 1988 ZL2BGJ had worked WA4NJP on 6m EME but he used a wire rhombic antenna and I had not considered that a single Yagi would work. At that stage Kjell had an array of four Yagis for Six. We ran many skeds and often we could receive part contacts, but the first of several completed contacts was on 5 January 2001.

In about 2001 Joe Taylor, K1JT, produced several computer programs, one of which is suitable for weak signal work and in particular EME communications. The mode JT65A has been widely used throughout the world and has made 6m EME contacts possible with relatively small stations at both ends, i.e. single Yagis with good ground gain.

The amount of ground gain, which may be up to 6dB, is dependent on foreground ground conductivity. The maximum antenna gain lobe is typically at a small elevation to the horizon and this is dependent upon the height of the Yagi above the ground. With a single Yagi there will typically be minor antenna lobes as well. For example, my 10-element Yagi at a height of 25m has a main antenna lobe at about 3°. However, there are several other minor antenna lobes up to about 18.5°. When the moon is at 3° I get maximum signal from the moon but I have worked big stations up to 18.5°. It is fortunate that often the moon is at low elevation in New Zealand and Europe at the same time. That is, moonset in New Zealand is often moonrise in Europe. It should be appreciated that a single Yagi with ground gain and no elevation is similar gain-wise to four similar stacked Yagis with elevation where there is no ground gain.

A good example of ground gain is demonstrated by working David, MM0AMW, with a single Yagi. There are both moonrise and moonset widows between David's QTH and mine. At David's moonset he is looking over the Atlantic Ocean and at his moonrise he is looking over land. That is, there are two EME windows to work New Zealand and while his moonrise window is typically longer it is far more difficult to complete an EME QSO. I have worked David both ways but his moonset over the ocean offers far better signals and I believe this is due to the superior reflection over the ocean. I have also had several great 6m EME contacts with Jan, OY3JE, with his single Yagi for similar reasons.

I have had many successful EME contacts after heavy rain or when there is snow on the ground and I believe this to be due to better ground conductivity and hence better ground gain.

The other requirement is low noise and often stations in city areas have problems with this, although I have worked several stations in cities, so it is possible. Often it will take several attempts and agreeing on a clear frequency to be used.

I have had many memorable 6m EME contacts but one was completing a 6m EME contact with Ian, M0BCG, now G5WQ. This was the first G – ZL 6m contact at a distance of 19,002km and occurred on 24 January 2005 at 1607UTC. Ian runs 4 x 5-element Yagis with elevation. I have had many contacts with Ian since that initial contact and it is appreciated when Ian often helps with monitoring and reports after my attempts with others smaller stations.

I have, however, noted the variation with which two stations quite close to each other may receive me. There are lots of variables with 6m EME, but one major one is Faraday rotation of signals and this can cause major problems should the signals be rotated close to 90°.

The greatest distance I have worked was with Joel, F6FHP, who had a single 9-element Yagi. The distance between us is 19,441km. At this distance, which is near the antipodes, the moon was at less than 4° during the common window, which was of just a few minutes duration.

It is interesting to note that although there has been 6 metre activity in New Zealand for many years, only Rod, ZL3NW, has achieved DXCC, with 103 to his credit at the time of writing.

REFERENCES

[1] ARRL DXCC 6m standings: https://tinyurl.com/y3u6ly83
[2] JI1CQA: http://ji1cqa.sakura.ne.jp/index.shtml
[3] JA1RJU (history, in English): https://tinyurl.com/y2ru5cm6
[4] http://zs6ez.org.za/lists/50zs1st.txt

Chapter 8

Portable and Expedition Operating

IF YOUR HOME location isn't ideal for 6m or 4m, one option is to head out to a suitable hilltop location and operate portable. You may do it simply for fun and as a change, to participate in one of the various portable contests that take place on the VHF bands, or maybe to try to catch some rare DX that you would be unlikely to work from home. The good news, for those of us in the higher latitudes, is that the main Sporadic E season occurs during our summer, when the weather is (hopefully!) more conducive to portable operation. Even if you don't intend to operate portable or head off on a DXpedition, this chapter ought to be of interest in giving some insights into what is required to mount such operations which you, from home, will enjoy chasing and working. Indeed, by stopping to think about what is involved from the 'DX end' of such an operation, you may well gain insights into how best to catch a DXpedition on the band.

Much of what has been written elsewhere in this book applies equally to portable operation, but there are other factors which you may need to take into account. It all depends on what you have in mind. Portable operation covers a multitude of scenarios, from backpacking to hilltops with a hand-held radio to mounting a full-scale DXpedition to an inaccessible, inhospitable island.

Your choice of location for a portable operation will be governed by a number of factors. If you want to work an announced DXpedition, for example, you may head for a seaside location, as the take-off over the sea gives a real boost to the low-angle component of your signal (lower loss in the near field of the antenna means that the important low-angle signals suffer less attenuation than over normal ground). If you want to operate a contest when conditions are likely to be average or flat, you may want to head for a hilltop location that is central to the various high levels of activity (surrounding towns and cities). In other words, you will need to give some thought to the ideal location for what you have in mind and then, ideally, make some reconnaissance beforehand to check factors such as accessibility, the need for permits (if any), facilities for camping if that is part of your plan, and so on. Every case will be different. But there are a number of factors which are common to most portable operations and these are discussed below.

PERMITS AND LICENSING

As this book is intended for an international audience it isn't possible to give specific details of what permissions are going to be required for portable operation. Everywhere will be different. But, generally speaking, anywhere that is privately owned will require you to have the permission of the landowner before you start setting up antennas. In the case of DXpedition operations from rare locations, those permissions may well need to be in written form to satisfy the various awards bodies that you were indeed where you claimed to be and with the necessary authorisations.

As far as licensing is concerned, if you are operating from your own DXCC entity there is usually no issue, though you may need to sign with some sort of portable designator (this is no longer mandatory in the UK, though it is required by the rules of some contests). There is a lot of material available about licensing in different countries (e.g. **[1]**), and the CEPT licensing arrangements have made it possible for many amateurs to operate from any one of over 100 DXCC entities that are now on the CEPT list.

This is the good news. The bad news is that 6m, and especially 4m, are not necessarily included in such arrangements, and it is necessary to check before firing up on the band and perhaps incurring the ire of local amateurs who, understandably, don't want to see their licence conditions treated in a cavalier fashion.

An example of the kind of situation that has to be dealt with is the 3B7C expedition of 2007, of which the author was a member. The standard Mauritian licence (3B6, 3B7 and 3B9 are legally parts of Mauritius though counting separately for DXCC) did not at that time include 6m and, indeed, the 3B7C licence, when it arrived, did not mention the band. Some negotiations with the Mauritian licensing authorities ensued and it became apparent that the reason for the lack of 6m authorisation was a concern that any operations would affect VHF TV in adjoining nations (other countries in the Indian Ocean region). In the end a compromise was reached, whereby three discrete frequencies were allocated to the expedition. This was fine, as these could be publicised to the DX world beforehand and there was, of course, no restriction on listening frequencies.

The 7-element 6m beam at 3B7C, St Brandon.

DXPEDITIONING ON 6M AND 4M

Some hardy 6m enthusiasts have been mounting 6m expeditions for years. Jimmy Treybig, W6JKV, and Lance Collister, W7GJ, spring to mind as classic examples. Once the number of amateurs active on Six started to increase, with the release of the band in many European and other locations, a 6m expedition became an interesting proposition, especially if planned to coincide with the best time of the year for long-haul propagation, usually June / July to take advantage of the main Sporadic E season in the northern hemisphere, where the majority of 6m DXers are located.

Once 6m was added to the ARRL DXCC Challenge as a qualifying band the demand for 6m operation during DXpeditions became even greater, as all those serious HF DXers suddenly needed a 6m contact to keep their place at the top of the listings. As a result, many HF-orientated DXpeditions now include 6m as a matter of course. However, the decision process regarding Six is likely to be different to a dedicated 6m DXpedition for a number of reasons, such as timing: HF DXpeditions are often scheduled around the equinoxes for best low-band propagation, but this is not optimal for 6m except perhaps at sunspot maxima. The best window for 6m EME may be different again, as discussed in Chapter 6.

LOCATION, LOCATION, LOCATION

It is notable that when well-known 6m DXpeditioners head off to a rare spot, they often seem to make their own propagation. The local stations in that area (parts of the Caribbean, for example) may rarely be heard on the band and yet, when these visitors show up, they seem to have regular band openings.

One of the main secrets to this is choice of location, though good antennas and plenty of power also help. The local amateurs will normally have their station at their home, which is probably in one of the local towns where they have their employment, where their children go to school, and so on. In other words, surprising as it may seem, amateur radio is not the main criterion for their choice of location!

They may also be constrained in terms of power by EMC problems. This is where the visitor has the advantage. He, or they, can select a suitable mountain-top or beach location with a clear take-off to one or more of the main areas of interest (North America, Europe and Japan) and, hopefully, no near neighbours. Cliff-top sites are perfect. Perhaps not the sort of place you would want to build your house, but ideal for a temporary expedition station.

There is nothing to beat a site visit to find a good location, but there is a lot you can do beforehand if you are planning any sort of DXpedition. Examples include contacting others who have been to the area before, maybe e-mailing a local amateur in the vicinity of where you want to go and, nowadays, taking a look on Google Earth which often has some great aerial shots which can give you a very good idea of what to expect.

BE QRV

As has been mentioned frequently in this book, 6m openings can be fleeting and if you have gone to the trouble of setting up a portable or short-term expedition station for whatever reason, you don't want to miss any of those openings. But this is where the short-term operation should have the advantage. The local amateurs will have jobs and families to think about. Even when they become aware of a band opening, they may have to get home from work and do the best they can from their less-than-optimal location. They may have some portable kit that they can throw into a car and drive to a hilltop, but by then the opening may well be over.

You, on the other hand, are there for the purpose of radio. You can be on or near the radio at an optimal location 24/7, at least for the duration of your expedition. Or, at least, this is what you should be planning to do, though there may well be distractions by way of sun, sand and entertainment.

The trick is to cater for this in some way so that the openings don't get missed. The usual solution is to set up and run a beacon transmitter for the duration of your expedition, so that DX chasers, many of whom may be looking out for you, will quickly become aware if the band opens. All they then need is some way to alert you to the opening, whether by texting your mobile phone, calling you on the beacon frequency between beacon transmissions or, in the case of a multiband expedition, calling you on HF to let you know.

A classic example is the 9M0C Spratly Island operation of 1998 in which the author took part. We left a 6m beacon running 24 hours a day. One day the beacon frequency came to life when a loud Japanese station called in the middle of the day, a clear sign that the band was open. Kazu, JA1RJU, the expedition's 6m operator, was in the swimming pool at the time, but

Kazu, JA1RJU, operating on 6m from the Spratly Islands, 9M0C.

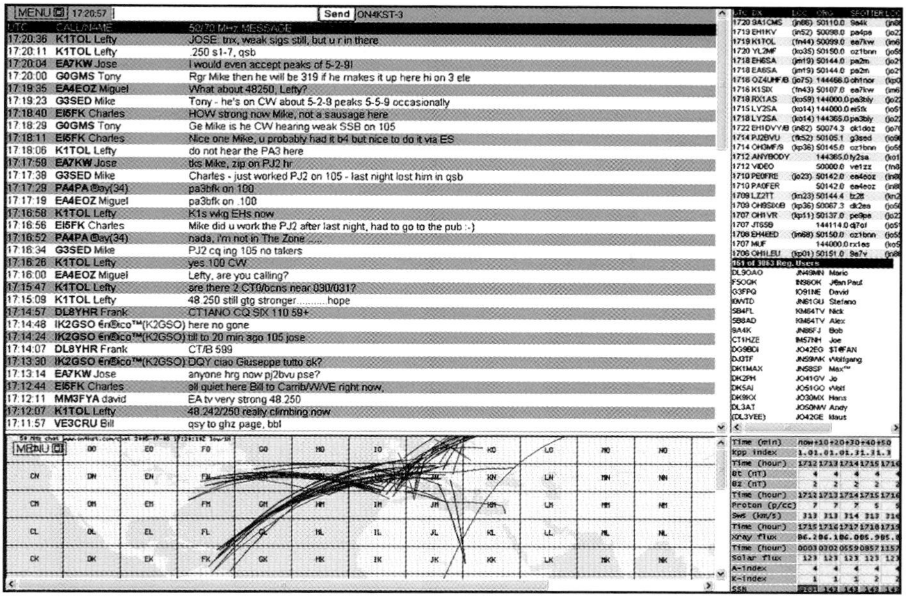

Screenshot from the ON4KST 6-metre chat room.

was quickly summoned by one of the HF operators who had heard the callers on 6m and Kazu arrived, dripping wet, ready to run the pile-up. Such is the dedication of the true 6m enthusiast!

COMMUNICATION

The previous section mentioned the importance of DXers being able to alert an expedition station to a band opening. Nowadays there are many ways of doing that, not only by amateur radio means but by way of the mobile phone system and the Internet. Expeditions to the more remote locations are increasingly taking satellite phones and, while not necessarily publicising the number to all and sundry, may, for example, nominate a friend back home to take calls and pass on any relevant alerts or feedback. There is also a lot to be said for having the Cluster network available at the expedition location and / or Internet facilities to connect to sites such as ON4KST chat, so that there is real-time feedback about what is happening on the band (to be able to watch propagation gradually moving towards you, for example). ON4KST is essential if 6m EME is included as part of the expedition plans. The good news is that mobile telephony has developed quickly even in the most out-of-the-way countries and locations so that good mobile Internet is often available at minimal cost (usually by buying a local SIM card).

WHAT FREQUENCIES & MODES TO CHOOSE

If your portable operation is for a contest, then choice of frequency and mode is a non-issue. You will simply go along with whatever is appropriate

A simple but effective 6m portable contest set-up. It's not always necessary to utilise a massive antenna array to achieve success on 4m or 6m, as is well illustrated by this photo of the G7DDN/P station. A home-made 3-element Spiderbeam modelled in *EZNEC* and fashioned from bits of wire and broken fishing poles was all that was necessary to win an RSGB 6m 'Backpackers' contest.

for the event. If, on the other hand, you are heading out on an expedition operation these are key considerations.

As a first example, suppose you are heading out to the Caribbean for a 6m expedition operation during the Sporadic E season. At times signals will, hopefully, be strong into the USA and possibly into parts of Europe. You will almost certainly be able to run stations faster on SSB than on CW and, at the same time, satisfy demand from those operators who are unable to operate CW. Unless your licence specifies spot frequencies you will need to choose one or more frequencies for your operation and, ideally, publicise them beforehand.

But other expeditions will almost certainly be heading out at this key time of the year, so a degree of coordination is desirable. Certainly, you cannot all expect to have access to the 50.110MHz calling frequency or chaos will ensue. So you have the option of trying to call CQ on that frequency and then announcing a move to another frequency or, as tends increasingly to be the case, announce a frequency before your departure and, unless there are good reasons to move (QRM on that frequency in certain target areas, clash with another operation etc), stick to it. This probably means using that frequency for all modes and expeditions are frequently to be heard on say, 50.103, running alternately on CW and SSB.

But signals won't be loud all the time. As has been mentioned elsewhere in this book, to make the most of Six you have to be prepared for true weak-signal working. In the past the most common weak-signal mode would have been CW, which has a significant signal-to-noise advantage over SSB when copied through a good quality narrow bandwidth filter. So, most expeditions would have been geared up for CW, with suitable equipment and, more importantly, competent CW operators who could run through a CW pile-up at a fast pace for when those short, low signal-strength openings occur.

A good example, during the 2008 Sporadic E season, was the TO5E St Barthelemy DXpedition which focussed on 6m (typically for such an operation, they had capability for the HF bands, to keep themselves amused when 6m was dead, but turned to 6m whenever there was the possibility of an opening). This activity made 1759 6m QSOs in 54 countries, a great effort, but 1435 of those contacts, a huge 82%, were on CW. In this case the operators chose to focus on CW, perhaps because of low signal strengths or maybe simply because they preferred that mode. Either way, as a DX chaser your best chance of a QSO was certainly on CW.

However, nowadays it is much more likely that a DXpedition will focus on FT8 for much of the time, because they will then be able to continue making QSOs even when band conditions are quite marginal. Yes, they may well move to SSB if the opening becomes sufficiently big and stable but, even then, they may be side-lining the lesser-equipped callers who, nevertheless, would have stood a chance on FT8. It can be a difficult call – QSO rates can be a lot higher on CW or SSB when the band is wide open but there is no doubt that using FT8 gives all callers a more equal chance of getting through.

But what about those times when there is no F-layer or Sporadic E opening? Do you sit by the pool and relax, or do you try other modes of operation?

This is a particular challenge for the HF expedition that also wants to operate Six, as the expedition may be taking place well away from the usual Sporadic E season. There are still options available, as for working from home, by way of meteor scatter and EME. Meteor scatter and other scatter modes can be useful if you are in scatter range of at least one of your major target areas. So, MS would be a good option for a station in the Mediterranean wanting to work around Europe or a station in the Caribbean wanting to work the USA, as an example.

But if you are in a more remote location EME may be the only viable option. Setting up a 6m EME capability during an expedition is far from trivial. It's hard enough from a home location. But it has been done and to

ZL8R from the Kermadec Islands was primarily an HF expedition that used EME to provide contacts on 6m.

great effect, a notable example being the ZL8R Kermadec Island expedition in 2006, which was essentially an HF expedition from a very rare location, too far away from most major countries to rely on the more traditional propagation modes for 6m working. So the team chose to use EME and were able to make a number of QSOs by that mode, even into Europe, a very long haul indeed.

The FT5XO expedition to Kerguelen Island also showed up on 6m and succeeded in working into the USA, a fantastic catch on the band. And, since then, as has been mentioned in a previous chapter, there have been a number of others, particularly by Lance Collister, W7GJ. Lance has actually written an excellent primer on planning for a 6m EME expedition; it can found on (and downloaded from) his website **[2]**.

Another 6m EME DXpedition by Lance, W7GJ: this one to the Cocos (Keeling) Islands in September 2017.

PORTABLE & DXPEDITION OPERATING

How does operating technique when portable or on an expedition differ from that from your home location? Firstly, you are the station that others are chasing. So, your time should be spent CQing, not waiting to hear signals. Most 6m DXpeditions, as mentioned earlier, set up a beacon at the earliest opportunity and leave it running when they are not making contacts. The usual system is that the radio goes back to receive during breaks in the beacon transmission and there should always be an operator nearby, even if he is reading a book and enjoying the sunshine, to hear any callers during those receive times.

At times when propagation is expected, the DXpedition will almost certainly be more proactive, calling CQ in the direction where propagation is most likely, and listening hard for the first sign of callers.

Once stations start to call, the normal rules of DXpedition operating apply. If you (the expedition) have been calling CQ on the calling frequency of 50.110MHz, now is the time to vacate that frequency and move elsewhere, probably to whatever frequency you have announced in your pre-expedition publicity. Once the pile-up starts to build, you will want to work split, probably with a reasonably wide split as you may be very weak with callers whereas they may well be very loud with each other, so you don't want their signals to be splattering across yours at the distant end. Similar considerations apply on 4m.

In terms of the QSO format, keep it short and sweet, for two reasons. Firstly, the opening may be very short and you want to put as many QSOs in the log as possible in the time available. If it's 'ragchews' you are after, then you shouldn't be on the 'sharp end' of a DXpedition! And, secondly, even within the length of a single QSO there may be significant fading, so keep the transmissions short. There is generally no need to include QTH locator, QSL information etc. This is all readily available via the Cluster network, your website and other sources. So just the callsign and the report. But if QSB does lead to a possible 'no QSO' it is worth hanging in there for a little longer in case the propagation takes an upward turn again and you can complete. Otherwise you could have some very disappointed customers.

On FT8, things are simpler but if conditions are good, it may be worth considering using FT8 DXpedition mode or switching to FT4 to increase the QSO rate.

The better-equipped expeditions ensure they have an online log, updated regularly (at least once a day, so that the update is done before the next day's opening to the same area). That way DX chasers can see whether they are safely in the log and, if so, desist from calling again. But if they aren't, there is still (hopefully) an opportunity for them to try again. Bear in mind, though, that many DXers nowadays may well be looking for a QSO on SSB and / or CW once they have an FT8 QSO in the bag. There is no fundamental harm in this provided that, in doing so, they and you are not depriving others from making their first contact with you. This is a judgement

The VK9ML Mellish Reef expedition made 4298 QSOs on 6m, over 50% on SSB.

call only you can make, knowing how far you have already succeeded in working through the pile of potential callers from a given area.

Indeed, it is reasonable to ask how many 6m DXers there are around the world. There is no documented answer, but major DXpeditions at times when 6m is in good shape will typically make two or three thousand QSOs. The Daily DX website shows the current record for 6m to be held by VK9ML (Mellish Reef, 2002) with 4298 6m QSOs. It is not surprising that the record was set from that part of the world, as Japan has probably the highest number of 6m DXers. Over half that total was made on SSB, contrary to what I said earlier about the use of CW but a reflection of the fact that the DXpedition took place at the peak of the sunspot cycle when there would have been some excellent, strong-signal 6m openings (VK9ML even recorded 146 RTTY and 56 AM QSOs on 6m during that expedition. Unfortunately, their website does not show a breakdown by continent.)

PUBLICITY

If you are heading off on a DXpedition you will want to publicise it in advance so that those who need you will be warned and can arrange to be around during your operation. Publicising a DXpedition has become dramatically easier in recent years thanks to the Internet. Generally, all you will need to do is to e-mail the major bulletins and perhaps some specialist websites such as the UKSMG site. As far as bulletins go, the Daily DX, 425 DX News and OPDX bulletins along with the DX-World website are probably the most widely read and an e-mail to these will ensure that news of your trip permeates to all the major areas. The important thing, though, is to send out that news in time for it to be picked up by national amateur radio magazines, which means allowing a good six to eight weeks before your trip to cater for publication deadlines. But at least, unlike in years gone by, you do not have to write to each magazine individually.

The other major element of pre-, during- and post-expedition publicity is to set up a website specifically dedicated to the operation. This is almost mandatory these days. A good website serves many purposes, advising the

DX community of your plans (timings, frequencies, callsign, etc), perhaps showing propagation forecasts for the main target areas, allowing log lookups once the expedition is under way, maybe offering a facility for requesting QSLs and / or making a donation and so on. Some of these facilities require that the website be updated once the expedition is underway, something that may be possible from your DX location, or may require a 'pilot' station back home who can work you on HF, perhaps, or via phone or e-mail, and then update the website with the latest news.

DURING THE EXPEDITION

Assuming you have some sort of Internet capability while on the DXpedition, it is increasingly common to put out news and upload logs on a regular (at least daily) basis. Club Log **[3]** is often the vehicle of choice for such logs, and offers the facility to include an OQRS (Online QSL Request) button.

Whether you choose to receive e-mails from all and sundry is a different matter – dealing with them can become very time-consuming and detract from the main expedition effort. On the other hand, it might keep you occupied between band openings!

QSLING, LOTW ETC

There is nothing unique about QSLing on 6m or 4m. As on other bands, those who work your DXpedition may want a QSL card in addition to an LoTW credit. In designing the card, though, it is important that you include the QTH locator as this is something that matters to 6m DX chasers. There are no other major specific aspects to consider, but if your expedition was focused on 6m and / or 4m you may well want to include some information about the equipment, site, antenna(s), QSO statistics, etc.

There are fewer 6m and 4m DXers and DXpeditions who use Logbook of The World (LoTW) compared with HF operators although this is changing rapidly with the onset of FT8. LoTW use is growing all the time and even though many 6m DXers will want your card for their collection, they may still prefer to claim as many DXCC credits as possible via LoTW in order not to have to risk their precious 6m confirmations to the vagaries of the postal service. So, it is good practice to upload your expedition log to LoTW at the earliest opportunity. If you haven't used LoTW before, there are instructions on the ARRL website and Gary, ZL2IFB, has written a great *LoTW New User Guide* **[4]**.

But check before you head off as you may need to submit paperwork over and above the licence documentation to have your operation accredited, especially if you were operating from a particularly rare location or from somewhere that doesn't normally allow 6m operation. Again (see previous section) some expeditions have taken to uploading to LoTW while the expedition is still in progress. But the danger of offering to do this is that if, in the event, it becomes impossible, the world at large will inevitably become frustrated – always on an expedition, whether in this matter or others, set expectations which you have a realistic chance of meeting.

Whether you upload to other web-based confirmation systems such as eQSL is very much a matter of personal preference. eQSL does not have the in-built security systems that are integral to LoTW, but some DXers are happy to print a QSL from the eQSL database for their personal use, rather than have the expense of sending off for a more traditional QSL card.

EQUIPMENT CONSIDERATIONS

The chapter on equipment choice is a starting point but in a portable or expedition situation you may have restrictions in terms of the size or weight of equipment you are able to take, especially if you are flying to your destination. So while, for example, an FT-857 may not have the same RF performance as its bigger brother the FTdx101, it may be the more realistic radio for your expedition.

There is a particularly hard decision to make about whether or not to take a linear amplifier and, at the end of the day, this may prove impossible. But, given what has been said elsewhere about the low signal strengths on Six during marginal openings, a linear amplifier is undoubtedly an attractive 'extra' if there is any way that it can be taken along. It is no surprise that Chris Gare, G3WOS, has found the 6m DXpedition linear amplifier design on his website to be one of the most- visited pages **[5]**.

The necessity of a linear amplifier was well illustrated in June 2008 when Dennis Motschenbacher, K7BV, undertook a 6m DXpedition to Belize (V36M)

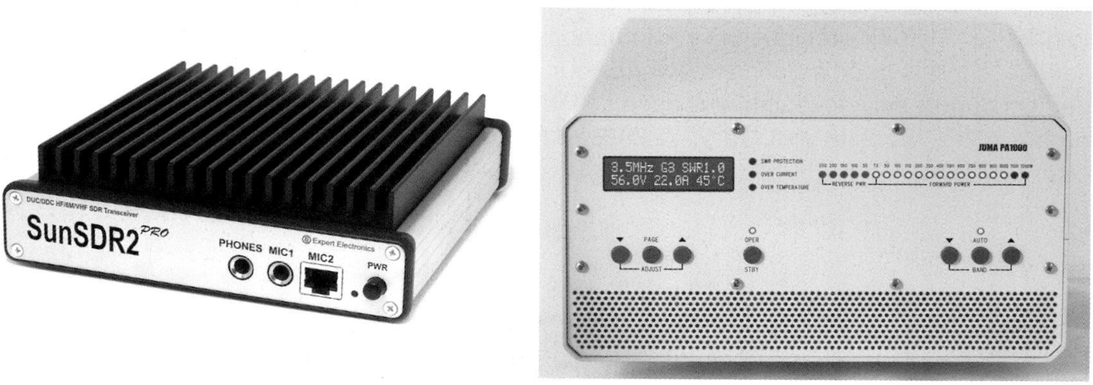

The SunSDR (10W) and Juma PA1000 amplifier make for a lightweight but effective 1kW station from HF to 6m.

and San Andres and Providencia (5J0M). His expedition amplifier went faulty shortly after his arrival in 5J0 and this undoubtedly had a major effect on the success of the DXpedition. For example, he worked very few UK stations, occasionally being heard at low strength. If the linear amplifier had still been operational, those openings would have been workable for significantly longer periods and at much better strength, allowing many more DX chasers to make it into the log. No doubt Dennis was very disappointed at what happened: with the best will in the world it's hard enough to carry one amplifier on a DXpedition and well-nigh impossible to take a back-up unit, so problems like this will inevitably occur from time to time. The good news is that, over the past few years, a number of lightweight amplifiers have become available from Ameritron, Elecraft, Expert, Juma and others which cover 6m at the 500 watt (or higher) level. Suitable amplifiers for EME are somewhat more of a challenge in that a power level of 800 watts or so is desirable (you will almost certainly be working with a single Yagi, albeit a substantial one) at a high duty cycle, so something like the SPE Expert 1.5K-FA would really be the minimum requirement. That particular amplifier weighs just 9.5kg, considerably lighter than its bigger brother the Expert 2K-FA but able to run similar power on data modes.

That said, KB7Q has operated from Curacao, PJ2T, with an Elecraft K3, KPA500 amplifier and a single Yagi, and successfully made EME QSOs. There is more information and advice on lightweight EME DXpeditioning on his website **[6]**.

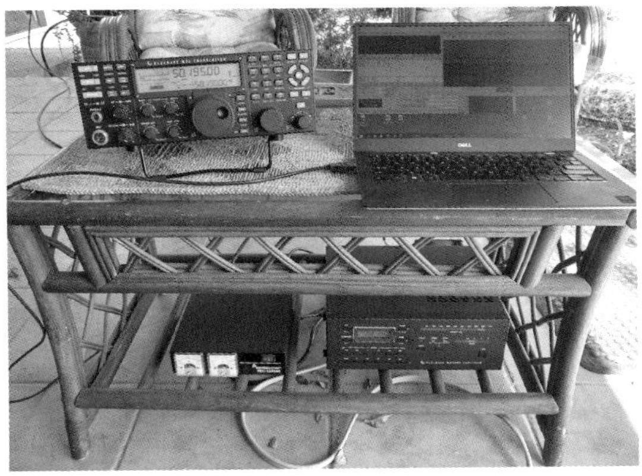

KB7Q made a successful 500-watt EME DXpedition to Curacao.

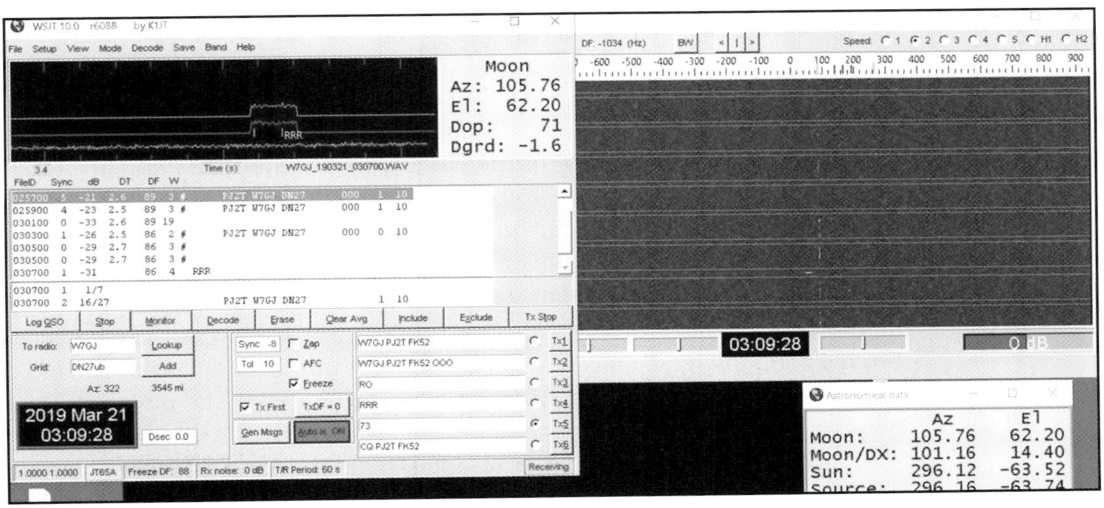

The EME QSO between PJ2T and W7GJ.

Keen VHF operator Bo Hansen, OZ2M, also operates on portable DXpeditions. He used a 4m 7-element Yagi to activate the Åland Islands as OH0/OZ2M.

Mention has been made of the desirability of running a beacon during a 6m DXpedition. This is not difficult to do. Many CW keyers have a beacon facility that is easily programmable. But most logging programs have a built-in CW capability which can also be programmed to beacon. Even the internal keyers of some transceivers can be set up to run a beacon.

As far as expedition antennas are concerned, again it comes down to what you are able to take. A key element of W6JKV's success on 6m expeditions has been the long-boom Yagi that he takes along. A good antenna not only gets out well but hears well, too, and may also be useful in nulling out local noise sources. Effort expended in getting a good antenna to your destination is generally effort well-spent.

The good news is that the hardware to get your antenna into the air can often be sourced locally once you arrive at your destination. Scaffold poles, ropes, pulleys and suchlike can be obtained from builders' merchants, ships' chandlers and similar stores which are found pretty much everywhere around the world. It may even be that local amateurs can help you out, too. The method of erecting a scaffold-pole mast using a gin-pole is shown in **Fig 8.1**. The mast typically consists of two or three scaffold poles (giving a height of 40ft or 60ft), with another pole for the gin-pole. The guys are 8mm polypropylene rope.

Coaxial cable may be harder to come by locally and this is something you may well need to check out before your trip. Hopefully you won't need too much. The sort of expedition that uses a scaffold mast attached to the

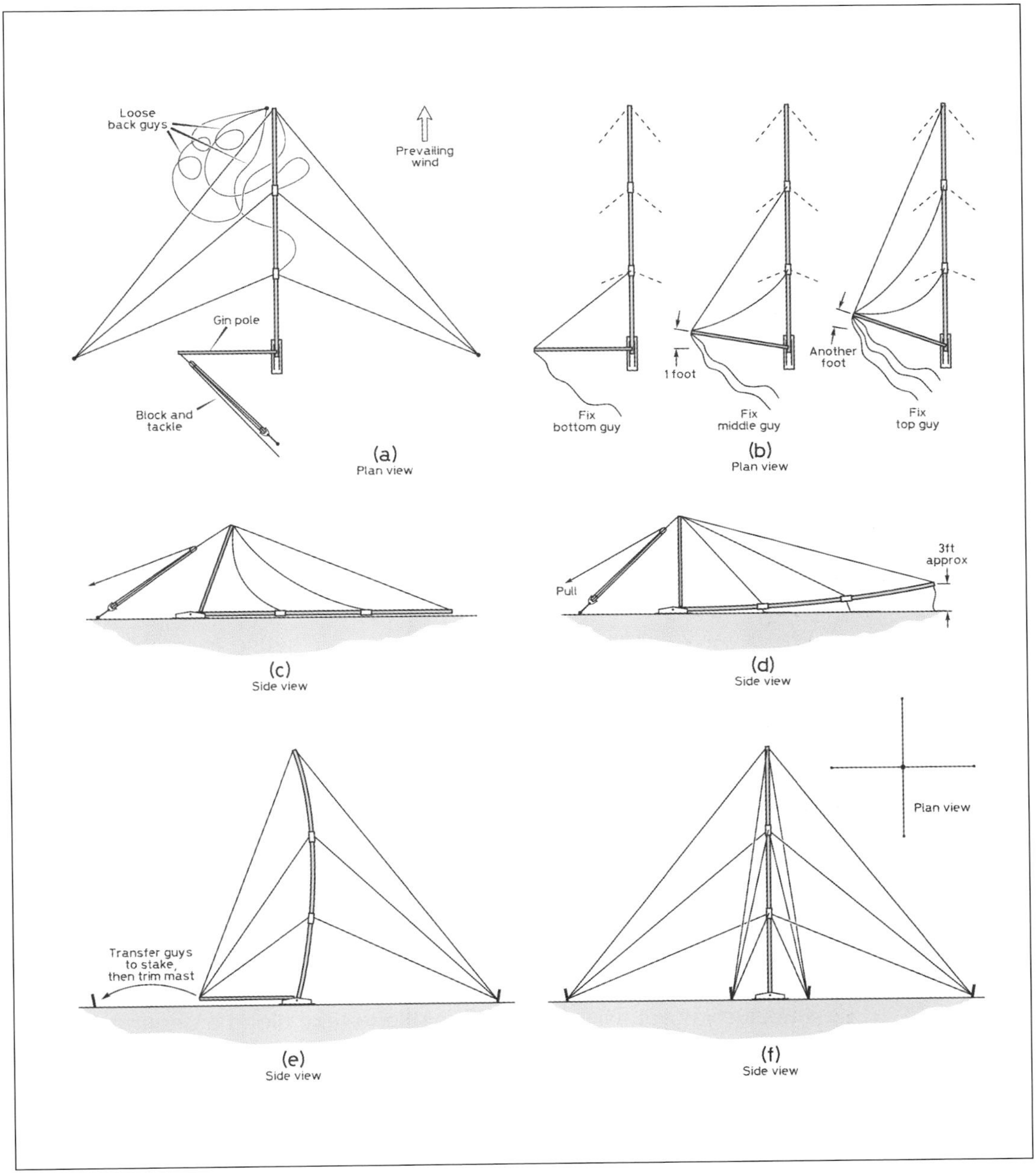

Fig 8.1: The gin-pole ('falling derrick') method of erecting a scaffold-pole mast. (a) Overhead view showing the mast and gin-pole laid out on the ground prior to erection. (b) Setting the pre-bend; the amount varies depending on the top load. (c) Side view showing the gin-pole being raised using the block and tackle. (d) As the guys take the strain, the mast should be slightly bent as shown. The antenna can now be fitted to the top of the mast. (e) With the mast in position, the gin-pole guys are transferred to stakes and adjusted to remove the bend (the other two sets of guys are not shown for clarity). (f) The mast in final position. Note the procedure should be reversed to lower the mast, ideally lowering away from the wind.

roof rack of a hire car which doubles as the shack, for example, may only need a few metres of feeder, with the antenna sitting directly above the operating position. Bigger expeditions will plan differently – the UK-based Five Star DXers Association, for example, generally took along a single long-boom Yagi or, occasionally, a stacked pair, which were put up on 40ft (13m) of scaffold poles and although the plans required the 6m antenna to be the nearest to the operating shack, the feeder run can easily be 100ft (30m) or so. But the team always tried to ship all their equipment in a 20ft shipping container, so that size and weight were not really an issue.

But do remember that coaxial cable becomes more lossy the higher the frequency. Multiband expeditions plan to have the 6m antenna system as the closest to the shack (shortest feeder run) and equip it with the best feeder they have available.

As for physically handling a large Yagi for 6m EME portable operation, various methods have been used to cater for the differing elevation angles required in order to track the moon. The T32C team, for example, were able to persuade the hotel's handyman to build a wooden support with hinge that allowed the Yagi to be tracked by pulling on and then tying down various ropes attached to the boom of the antenna. Somewhat 'Heath Robinson', but it worked!

Whatever you plan, it doesn't usually make sense to use an antenna rotator. The weight of a rotator makes a tilt-over mast (which you will probably be using, unless you have access to some sort of telescopic push-up or pneumatic mast) much more unwieldy. Usually it is simpler to design the system so that it can be rotated by hand from the bottom. Remember to take along a compass to check the beam headings and then use stones, sticks or some other marker system at the base of the antenna to remind you of the main beam directions for North America, Europe, Japan etc.

GENERATORS / POWER SOURCES

One of the key considerations for a portable operation is power. If you are on some sort of Field Day operation with your local club this is unlikely to be an issue, as you will probably be able to take along a generator in the back of a car. But shipping generators or heavy-duty batteries to the far side of the world is not to be recommended, unless they are part and parcel of a major DXpedition shipment. More likely you will want to source your power locally at your destination.

Generators can be hired almost anywhere in the world and modern units are reliable, provide a well-regulated supply and are often quite quiet too. Battery power is fine for a 100-watt rig – a good car or truck battery will last for several hours, or more if you can charge it from your hire car. But battery power is unlikely to be sufficient if, as has been recommended, you use a linear amplifier. Ideally, you can find a rental villa that is well sited and has mains electricity, though even that is not necessarily the perfect solution that it may seem, as electricity supplies can be notoriously intermittent in

Diesel or petrol generators can be more reliable and are often less electrically noisy than the mains supply in many countries. This Yamaha 2.5kVA petrol generator with both 110V and 220V AC outputs was old but nevertheless proved reliable and did not produce any electrical noise.

some DX countries, with power outages inevitably occurring just as that much-awaited band opening comes along. Noise levels from mains power lines can also be a real problem in many countries.

Be sure to check what voltages will be available locally. Fortunately, most modern equipment can be switched between 220 and 110V as required.

OTHER CONSIDERATIONS

Bear in mind other considerations such as health and safety issues, especially if your trip takes you a long way from the nearest medical facilities. They also include the need to take plenty of spares (connectors and suchlike) plus simple tools and test equipment (soldering facilities, multimeter, antenna test meter, for example), especially as you are likely to end up a long way from the nearest emporium. It isn't easy, especially if you have travelled by air, with the ensuing weight and other baggage restrictions and compromises will have to be made. But it must be so frustrating to be in the position that a failure occurs, such as the 5J0M problems mentioned earlier, and not to have the tools to hand to make at least an attempt at a repair. Travel is actually one of the big limitations nowadays, in this age of worries about terrorism and, in 2020 – 2021, the Covid-19 pandemic. Many officials are unfamiliar with amateur radio gear and may be suspicious of what you are carrying. Plenty of advice exists about how to handle this and it is worth thinking through beforehand such issues as what to carry in your hand baggage and what to check in (with the risk that it may arrive late at your destination or even go completely astray). If nothing else, do have your amateur radio licence to hand, along with any other relevant documentation, as proof of what it is that you are about.

DXPEDITION FUNDING

Finally in this chapter a quick word about expedition funding. DXpeditions will never be a way of making money, despite disparaging comments that occasionally appear on some Internet reflectors. They are expensive to mount and the only direct source of 'income', that is to say contributions sent with QSLs, is unlikely even to cover the cost of printing and posting the cards, especially if your expedition has been confined to one or two VHF bands, so that QSO totals are relatively modest.

If you consider the DXpedition as simply a form of activity holiday which you are happy to fund, then that is fine. But it is worth being aware that there are groups and foundations that are in a position to help with some of the radio-specific expenses of a legitimate expedition to a suitably rare location. You will be expected to cover your travel and accommodation costs, but bodies such as the UKSMG **[7]** or the Northern California DX Foundation (NCDXF) **[8]** are in a position, if a suitable case is made, to help with certain expenses, such as the cost of shipping an amplifier or antenna beforehand, or maybe offering the loan of a suitable lightweight transceiver.

Whether you are asking for money or not, it is often worthwhile to contact such bodies before your trip, to find out other information about matters such as licensing, the rarity of your proposed location etc. They are often in a position to give useful advice.

REFERENCES

[1] *World Licensing and Operating Directory*, by Steve Telenius-Lowe, 9M6DXX, RSGB, 2008.

[2] 'QSO Procedure and Tips for this and Future Operations', Lance Collister, W7GJ: https://tinyurl.com/p7lpz9o

[3] Club Log: www.clublog.org

[4] ZL2IFB's *LoTW New User Guide*: https://www.g4ifb.com/LoTW_New_User_Guide.pdf

[5] G3WOS 6m amplifier: www.gare.co.uk/amplifier

[6] KB7Q VHF / UHF DXpeditions: https://tinyurl.com/y2u26mwa

[7] UK Six Metre Group: www.uksmg.org

[8] NCDXF, PO Box 2012, Cupertino, CA 95015-2012, USA; www.ncdxf.org

Chapter 9

Contests, Awards and QSLing

THIS CHAPTER COVERS a miscellany of topics that have been touched on elsewhere in the book and which will add to your enjoyment of Six and Four. Firstly, it covers some of the major contests on those bands. Whether or not you consider yourself a contester, contests generate activity and are therefore a great way to put contacts in your log including, all being well, some new squares, counties, or whatever it is you like to chase. And that takes us nicely on to the second topic, awards.

Because, whatever you do like to chase, there will almost certainly be an award that covers it, whether it be squares, countries, islands or states. There is a lot of information on the Internet and elsewhere but this chapter covers some of the most popular operating awards as they relate to 6m and 4m.

If you want to chase awards, you will need to collect QSL cards and / or participate in the ARRL's Logbook of The World (LoTW) database, so the final part of the chapter covers these topics, albeit in fairly short order as there is plenty of material available from other sources.

CONTESTS

Contest operating has been mentioned already, in Chapter 5. Contests are an excellent opportunity to learn about propagation, to add to your totals, to build your operating skills and, of course, to have fun. One of the problems with 6m and 4m propagation, notwithstanding the excellent beacons that exist, is that the bands can be open somewhere from your QTH and you may be totally unaware of the fact because there is no activity at the far end. During a contest there are much greater levels of activity, giving a far better idea of band openings.

There are quite a number of contests that cover both bands, either as specifically 6m or 4m events or as part of a multiband contest. Most take place over the (northern hemisphere) summer period. 6m and 4m events are scheduled at that time because Sporadic E propagation is more likely to be about, making the contest more interesting. Multiband VHF events are scheduled for the summer because the higher VHF and UHF bands are essentially line-of-sight (most of the time) so portable operation from elevated locations is encouraged and it's not much fun setting up on mountaintops in the middle of winter!

The following paragraphs mention some of the scheduled 6m and 4m events in Europe and North America. These are subject to change and you should certainly check dates and rules from the relevant websites before packing your camping gear for the weekend or arranging for the XYL to be away so that you can do a serious contest effort from home.

In the UK, the main organisers of 6m contests are the RSGB and the UKSMG. The UKSMG Annual Summer Sporadic E Contest takes place in early June and is popular throughout Europe. The rules can be found on the UKSMG website and you don't need to be a UKSMG member to take part.

There are several RSGB contests that are of interest to both 6m and 4m operators. These include the following:

50MHz UK Activity Contest (every second Thursday)
70MHz UK Activity Contest (every third Thursday)
50MHz MGM Activity Contest (every fifth Tuesday)
First 70MHz Contest (April)
First 50MHz Contest (April)
70MHz CW Contest (May)
50MHz CW Contest (June)
VHF National Field Day (July)
70MHz Trophy (July)
70MHz AFS Contest (September)
50MHz AFS Contest (October)
50 / 70 / 144 / 432MHz Christmas Cumulatives (December)

The dates, rules, etc are published in the *RSGB Yearbook* **[1]** and on the RSGB VHF Contest website **[2]**.

The CQ World Wide VHF Contest is held in July and includes 6m. This contest, although most popular in North America, also generates quite a lot of activity in Europe. Details from the CQ website **[3]**.

The SMIRK QSO Party, sponsored by the Six Meter International Radio Klub, is held in June of each year. You do not need to be a SMIRK member to participate. Details from the SMIRK website **[4]**.

The ARRL runs the following events that include 6m:

January VHF Contest
June VHF Contest
June Field Day
September VHF Contest
November EME (50 to 1296MHz)

Details, not surprisingly, from the ARRL website **[5]**.

The Four Metres Website lists almost thirty 70MHz events, several of which are local events (e.g. based in Scandinavia) although there is always the possibility that propagation may be favourable and contacts might be made with other areas.

There are no doubt quite a few other 6m and 4m contests around the world and if your national society doesn't sponsor anything appropriate, you might suggest that it does so. Here are a few others: the OZ 50MHz Open Cumulative Contest runs on the fourth Tuesday every month **[6]**, and the SARL (South African Radio League) run several contests which include 50MHz **[7]**. The UK magazine *Practical Wireless* also sponsors an annual 70MHz contest (in September), which remains popular, especially given the upturn in 4m activity in recent years.

In most respects contesting on Six and Four is no different to contesting on other bands. If you are taking part in the contest simply to find DX such as new squares and countries by way of all that contest activity, then all you need to do is tune the band and work them or, if you prefer, use the Internet-based facilities that have been discussed elsewhere such as the DX Summit website and the ON4KST chat room to be alerted to any particularly interesting contest activity. You may also want to run a local CW Skimmer or log into the Reverse Beacon Network. Be aware that, if you intend putting in an entry, the use of external spotting facilities may well put you in an Assisted or Multi-Operator category (some contests rule out such assistance completely, which would then require you to submit as a checklog).

Before working any contest stations, at the very least check the contest exchange (often simply RS(T) plus locator square, but sometimes including a serial number, UKSMG membership number or other data) so that, when you work someone, you know immediately what he needs from you. You can probably determine the exchange from listening to a few QSOs, but the best way is to check out the rules via the organiser's website. Any rare stations in the contest (particularly if they are out portable to activate an unusual grid square) will want to amass as many contest QSOs as possible, so contests are not the time for exchanging names or other extraneous information.

If you plan to take part seriously in contests, then rather more effort is required, both before and during the contest. Beforehand, check the rules carefully and plan your strategy based on them, especially the scoring system. Does it favour distance, multipliers (e.g. grid squares) or some other aspect? Check out your station, too, to ensure that there are unlikely to be any technical problems while the contest is underway. There's quite a bit more you can do, too, such as preloading the standard messages into your voice or CW keyer. Warn the family that you will be unavailable, think about your food and sleep strategies, and so on. All of this is familiar to contesters, whatever the band. If you plan to operate portable, maybe to take advantage of a hilltop location, then there is plenty of other preparation to do, covered in Chapter 8. Don't forget, for example, to check the grid

The single-operator 6m contest station of Roger, G4BVY/P, in an RSGB 50MHz contest. The station took about 30 minutes to set up and consisted of an Elecraft K3 transceiver running from the Land Rover battery to an F9FT 5-element Yagi, showing that a simple station can be easily set up while also being reasonably competitive. The location was a hilltop about 10km from Malvern.

square of your portable location and print it up on a card so that you don't accidentally send your home station grid during the contest! Check also, via the various bulletins or websites, what portable and other expedition activity will be taking place during the contest, so that you know what to be on the lookout for.

So how about aspects that are specific to Six and Four? Having read this far, you will be fully aware that these bands are noted for the many propagation modes that they enjoy. You will want to think about what propagation is likely during the contest (many VHF contests are, of course, scheduled for the Sporadic E season) and be ready to make the most of any band openings that occur. If, for example, the contest scoring is distance-based, you simply cannot afford to miss any fleeting openings even if it means passing up on a nice run of more local contacts. And if, say, there is a multiplier system based on squares and / or DXCC entities, then you will want to ensure that these go into the bag even if, again, you have to take time out from a run of more mundane stations to do so.

It's your choice whether you use Internet-based assistance or not. As mentioned, some contests allow it, some allow it but put you in a different category, some won't allow it at all, in which case you can still use such assistance but if you send in your log, it will be treated as a checklog. But there is no restriction on local alerting techniques such as a separate receiver monitoring any commercial frequencies that might alert you to possible band openings or the use of a radio with a spectrum display so that, while working stations on your frequency, you can get an idea of activity elsewhere in the band.

If the contest runs for, say, 24 hours, you will want to schedule any rest periods at times when other stations are least likely to be active and when propagation is likely to be poorest. This will almost certainly mean the middle of the night, when you are sure that any Sporadic E has finally died away.

There is no magic secret to becoming a successful contest operator, except to take part in as many events as possible so that the actual operating becomes second-nature and your brain can focus on the strategic aspects such as when to check different directions, when to sweep the band and when to CQ or when to switch modes (in multimode events). And during the quiet times you may even find it profitable to drum up some contacts from local stations operating FM simplex (repeater contacts are unlikely to be valid for contest points). Every contest is different and this is what makes them fun. You may be plodding away at a low QSO rate and getting frustrated, ready to take a coffee break just at the moment that the band explodes with Sporadic E propagation and your log starts to fill up with distant callsigns. When rates are slow you may well be happy to chew the fat a little with other stations you work, just to stay awake and alert. When it gets busy, the good operator will be ready to get his contest exchange away in a snappy fashion and copy the other station's exchange first time, without needing to ask for repeats.

Whether you record the whole contest to your hard disc to be able to listen again later is up to you. Some contesters consider this against the spirit of contesting while others would not be so concerned. The general consensus is that it is fine to make a recording for later analysis and interest, but not appropriate to use it for amending the contest log before submission.

To summarise, the main advice is to check out the rules beforehand, and think carefully about which category you intend to enter, your strategy to achieve a successful result, etc. What you consider a successful result may vary. You may be interested in using the contest to add to your total of squares worked, in which case you may choose to use 'assistance' during the contest, by way of Cluster or the ON4KST chat room. But, as I said earlier, some contest organisers will then only accept your log as a checklog. This is fine if you have met your goals, but if you want to make a serious entry check the rules first and if, for example, those facilities are ruled out for competitive entrants bear that in mind in your planning. If you are heading out portable for a contest the usual pre-planning is necessary.

And do send in your log afterwards, whether as a competitive entry or as a checklog (in the format specified in the rules, please). It will be appreciated by the contest organisers as a way of checking other entrants' logs and the contest database as a whole may even be helpful for later analysis as regards propagation mechanisms at work over the contest weekend.

AWARD CHASING

Award chasing is a popular pastime on all bands. Many of the major awards allow single-band endorsements or have a specific 6m or 4m version of the award.

RSGB 6m AND 4m AWARDS

The RSGB issues 50MHz and 70MHz Countries and Squares awards, the latter as part of the '4-2-70' awards programme. Further details are shown in the two sidebars and are also available in the *RSGB Yearbook* as well as on the VHF awards page of the RSGB website **[8]**.

RSGB 50MHZ AWARDS

There are three awards for the 50MHz (6m) band. They encourage you to develop your operating skills and station to achieve the most from what can be an interesting, yet challenging, area of the band plan.

50MHz 2 Way Countries Award

May be claimed by any licensed amateur who can show that they contacted amateur radio stations in the required number of countries. The basic level of achievement is 10 countries. Further certificates may be applied for in increments of 10 countries worked.

Rules: All contacts were made on or after 1 June 1987 from the same DXCC country. Only those countries formally authorising operation within the 50MHz band will qualify. Award applications will not be accepted for mixed categories. Only Fixed, Temporary or Portable stations count.

50MHz DX Countries Award

May be claimed by any licensed amateur who can show that they contacted amateur radio stations in the required number of countries. The basic level of achievement is 25 countries. Further certificates may be applied for in increments of 25 countries worked.

Rules: All contacts were made on or after 1 June 1987 from the same DXCC country. Only those countries formally authorising operation within the 50MHz band will qualify. Award applications will not be accepted for mixed categories. Only Fixed, Temporary or Portable stations count.

50MHz Squares Award

May be claimed by any licensed amateur who can show that they contacted amateur radio stations in the required number of locator squares. The basic level of achievement is 25 locator squares. Further certificates may be applied for in increments of 25 locator squares worked.

Rules: All contacts were made on or after 1 June 1987 from the same DXCC country. Only those countries formally authorising operation within the 50MHz band will qualify. Award applications will not be accepted for mixed categories. Only Fixed, Temporary or Portable stations count.

RSGB 70MHz AWARD

The RSGB 70MHz (4m band) award may be claimed by any licensed amateur who can show that they contacted amateur radio stations within the 4m band. There are 13 levels of achievement:

70MHz – 20/4 (20 squares and four countries)
70MHz – 25/6
70MHz – 30/8
70MHz – 35/8
70MHz – 35/10
70MHz – 40/8
70MHz – 40/10
70MHz – 45/8
70MHz – 45/10
70MHz – 50/8
70MHz – 50/10
70MHz – 55/8
70MHz – 55/10

Rules: All contacts must have been made on or after 31 December 1978 from the same DXCC entity. Eligible countries are those listed by the ITU. Award applications are eligible in the following categories: Fixed, Temporary or Portable stations.

The RSGB VHF awards for 50MHz and 70MHz.

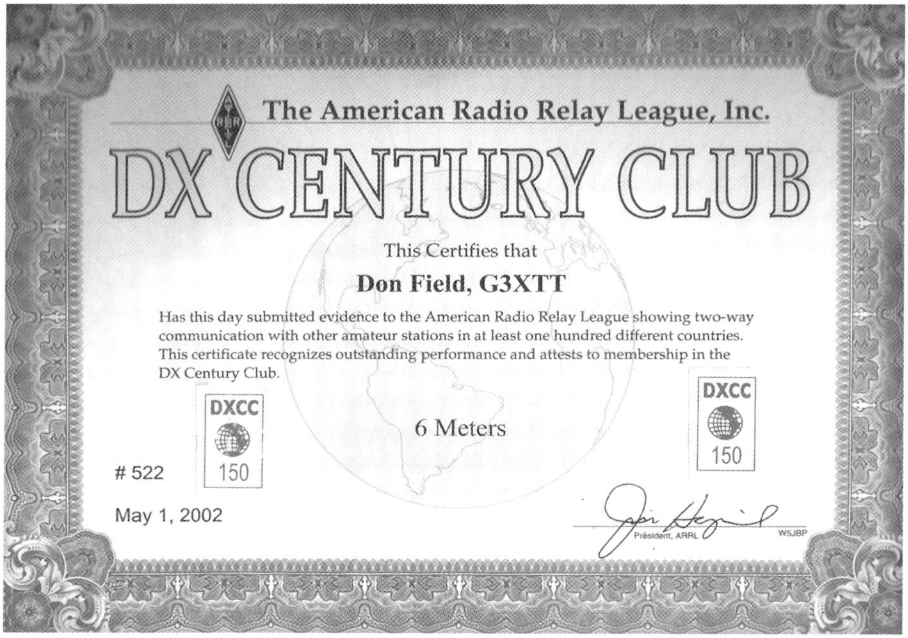

The author's 6m DXCC certificate, with endorsements for 125 and 150 entities.

ARRL AWARDS

DX Century Club (DXCC)

The 6m version of the ARRL DXCC (DX Century Club) award was introduced in 1990 and eight amateurs qualified that year (contacts from 1945 onwards were eligible). At the time of writing, over 1500 amateurs had received this award, which is for working and confirming a minimum of 100 DXCC entities from the overall list of 340. This is a tough achievement if you live somewhere like the US West Coast, whereas European amateurs have a head start, with something like 70 DXCC entities on their doorstep or, at least, within annual Sporadic E range. Even so, the number has increased by more than 500 since the last edition of this book, despite the recent lack of ionospheric propagation on 50MHz.

The DXCC Challenge plaque is available to those who have confirmed at least 1000 current DXCC credits on HF plus 6 metres.

Full details of how to claim DXCC are available from the ARRL website. The good news is that, since the advent of Logbook of the World you may not need to collect traditional QSL cards for the award application. If the stations you work submit their logs to LoTW, you can claim credits electronically. You may still want to collect QSL cards for the shack wall, but at least you don't have to risk sending them through the mail for checking.

A lists of early recipients of the 6m DXCC can be found

on ZS6EZ's website **[9]** and the current list of all holders, with their DXCC standings, can be downloaded as a PDF file from the ARRL website **[10]**.

An extension to DXCC is the ARRL 'DXCC Challenge' award, which is earned by working and confirming at least 1000 DXCC band-points, using any bands from 160 to 6m (except 60m). Those with more than 1000 band-points can claim a handsome wooden plaque, which is endorseable in increments of 500. The inclusion of 6m in the DXCC Challenge has encouraged many HF DXers to take a keen interest in Six too!

VUCC

The VHF / UHF Century Club award (VUCC) is another ARRL award, specifically geared to VHF operation. VHF / UHF operators have tended to want to chase some sort of geographical entity smaller than a DXCC country, because high country totals are tough on VHF, especially on 2m and above. The popular answer is the chasing of Locator Squares which were introduced originally as a way of pinning down station location for the scoring of VHF contests.

The original five-character QRA locator system, based on latitude and longitude, has its origins in Europe back in 1959, but was unsuitable for world-wide use because grid references would not be unique but would repeat at intervals. Therefore, in 1980, a VHF working group meeting in Maidenhead, UK, adopted a proposal by John Morris, G4ANB, based on a newly generated world-wide system devised by Folke Råsvall, SM5AGM.

The VUCC award, available from the ARRL.

This system covers the world in grid locators which are 2 degrees of longitude in width and 1 degree of latitude in height. If this seems odd it is useful to think that, at the equator, they will be twice as wide as they are high, but at the poles their width will be zero! At the latitudes where the bulk of the amateur population resides, they are not far off square (this system actually makes grid square chasing in, say South Africa, much tougher than in, say, northern Europe, because each locator 'square' actually has a much greater area in ZS). These basic grid squares are designated by two letters and two numbers (e.g. IO91), though a further two letters can be added to give a more precise location. More information on the Locator system can be found in Appendix G.

Locator square chasing is very popular on all VHF / UHF bands, not least on 6m. The rules of the VUCC award appear on the ARRL website. The basic requirement for a 6m award is 100 credits, which means 100 confirmed squares worked since 1 January 1983. In reality, from pretty much anywhere in Europe or North America, 100 grids should be workable in a matter of months, or even weeks, even with a very modest station, especially if you focus on those periods of the year when Sporadic E is common.

Worked All States

As well as DXCC, LoTW also supports the Worked All States award. WAS is certainly an achievable goal for North American and Caribbean stations active on 6m, but somewhat more challenging from, say, Europe or Japan. Single-band WAS awards are available for 6m operation. Application and record forms are available from the ARRL website.

The Worked All States (WAS) award.

Fred Fish Memorial Award (FFMA)

In 2008 the ARRL Board of Directors approved a new award honouring the late Fred Fish, W5FF, the only amateur who worked and confirmed all 488 grid squares in the 48 contiguous United States on 6 metres (and also the second recipient of 6m DXCC).

The ARRL press release announcing the award states, "The Fred Fish Memorial Award will be granted to any amateur who duplicates Fish's accomplishment. Fish was a mainstay on the VHF+ bands for many years, having achieved Worked All States (WAS) on 6 metres through 432MHz, as well as DXCC for 6 metres. He is widely regarded as a gentleman operator and one of the finest amateurs in the VHF+ community. ARRL Contest Manager Sean Kutzko, KX9X, will oversee this award. Kutzko, an avid VHF+ operator himself, said, 'We hope the new award will increase 6 metre activity throughout the US and the world. We also hope it will lead to the activation of rare grid squares in the US by encouraging the native ham population of a rare grid square to give 6 metres a try, as well as through so-called Grid DXpeditions. We actively call on the 6 metre community to help educate VHF+ newcomers to the fun that is available on 6 metres'." Complete details on the Fred Fish Memorial Award are available from the ARRL website.

Incidentally, for all ARRL awards amateurs in the USA and US possessions need to be ARRL members. Amateurs elsewhere in the world may apply without being ARRL members.

OTHER AWARDS

CQ 50MHZ WAZ

CQ Magazine issues the Worked All Zones awards for contacts with the 40 zones into which the world is divided for purposes of this award. A single-band award is available for 6m operation, Mixed Mode only, and is issued for a minimum of 25 zones confirmed, with endorsement stickers for 30, 35, 36, 37, 38, 39 and 40 zones. Contacts must have been made since 1 January 1973.

A list of 'top contenders' showing zones still needed appears on the CQ awards website **[11]**. At the time of writing LZ2CC, IW9HII and JE1BMJ have all 40 zones, while JH0BBE needs only zone 34.

UKSMG Awards

The UKSMG issues a number of awards for 6m operation. There are continental awards, for example for working 10 DXCC countries in the continent of Asia, a Digital award for working grid squares using one of the digital modes and various awards for increasing levels of achievement on 6m EME. Full details available from the UKSMG website **[12]**.

Islands on the Air (IOTA)

The Islands on the Air (IOTA) awards programme has a VHF / UHF version,

The IOTA award is becoming increasingly popular with 6m operators.

which is increasingly popular with 6m operators. Full details are available from the IOTA website **[13]**.

* * * * *

The above list of awards is by no means intended to be comprehensive and, in any case, many general operating awards can be obtained with a single-band endorsement for 6m. For a comprehensive source of information about amateur radio awards, the best single resource is the K1BV website **[14]**.

QSLING

Some amateur radio awards work on an 'honour' basis; in other words, your application simply needs a signed declaration that you have worked the stations concerned (or that you have confirmations of those contacts). Some require a third party to see your cards and endorse your application (typically the rules may require that your application is endorsed by two officials of a local radio club). Others require that you send QSL cards to the organisation that actually issues the award.

Whatever the circumstances, and even if you don't chase awards, it is always nice to have a QSL card as a permanent record of a memorable contact. The arrival of the QSL is also reassuring if the contact was made under marginal circumstances and, perhaps, you weren't 100% sure that the contact was OK!

The QSL bureau system has worked effectively for many years, but it can take a matter of years for your card to reach the distant end and a card to find its way back to you. That's fine for the bulk of contacts that you

Peter Lund, LA7QIA, first operated on 6m as JW7QIA in 1996. In 2010 he activated 4m for the first time from Svalbard. A superb QSL on 6m or 4m in anyone's collection!

make in day-to-day operation or in a contest, but for that rare one that you need for an award or to put on the shack wall you will want a somewhat quicker turnaround.

Within the USA the ARRL bureau only handles cards for overseas contacts, so any domestic cards American operators need for, say, WAS, need to be chased directly via the postal service. Most DX operations and DXpeditions appoint a QSL manager to speed up the process and, in some cases, to get over the problem of postal theft that is common in many countries.

Lots of guidance has been offered on how to do direct QSLing to maximise your chances of a return card and I won't repeat it here. Check sources such as the *RSGB Operating Manual* for detailed advice. But generally it is a matter of common sense, such as including an addressed return envelope with funds for return postage (usually a couple of dollar bills given that International Reply Coupons are now effectively a thing of the past) along with your QSL card on which you have correctly entered all the QSO information (remember, for example, to put times in UTC, not local). And if sending overseas, it's generally wise not to put any reference to amateur radio on the outside of the envelope or any suggestion that it may contain something of value.

If you are planning on having QSL cards printed, ensure that your card carries information that is of importance to other 6m and 4m operators, i.e. your country, state (in the USA) or county (in the UK) and, most particularly, your grid square. You may also want to add equipment and antenna details: many 6m enthusiasts have two designs of card printed; one specifically for their 6m QSOs and one for everything else!

LOGBOOK OF THE WORLD

The ARRL's Logbook of the World (LoTW) is now well-established and supports the DXCC and WAS awards programmes. You can apply for DXCC awards and endorsements using a mixture of traditional QSL cards and matches on LoTW. The latter requires that you upload your own log to LoTW and the system will match any contacts it is able to with other logs on the LoTW system.

Increasingly, DXpeditions are uploading their logs to LoTW as a matter of course, so this can be a convenient and inexpensive way of getting those credits for DXCC. The ARRL has enhanced the LoTW software to enable links to other popular awards programmes (for example, *CQ Magazine's* WPX awards and the IOTA awards programme are now supported), but this is by no means trivial while protecting the integrity of the LoTW database, so don't expect further immediate developments on this front.

eQSL

While on the topic of QSO databases for awards purposes, it is worth mentioning that there are other such databases in existence, the most popular of which is eQSL **[15]**. eQSL does not have the wraparound of public key encryption that is fundamental to LoTW and is therefore potentially more open to abuse. But if you simply to want to see matches to your contacts and print attractive cards for the shack without the cost of direct QSLing, you may want to give eQSL a try. It is certainly true that many amateurs nowadays (particularly of the younger generation) see no role for 'traditional' paper QSL cards and would prefer to have all their confirmations in electronic form with, perhaps, the option to print out a confirmation locally (which eQSL offers).

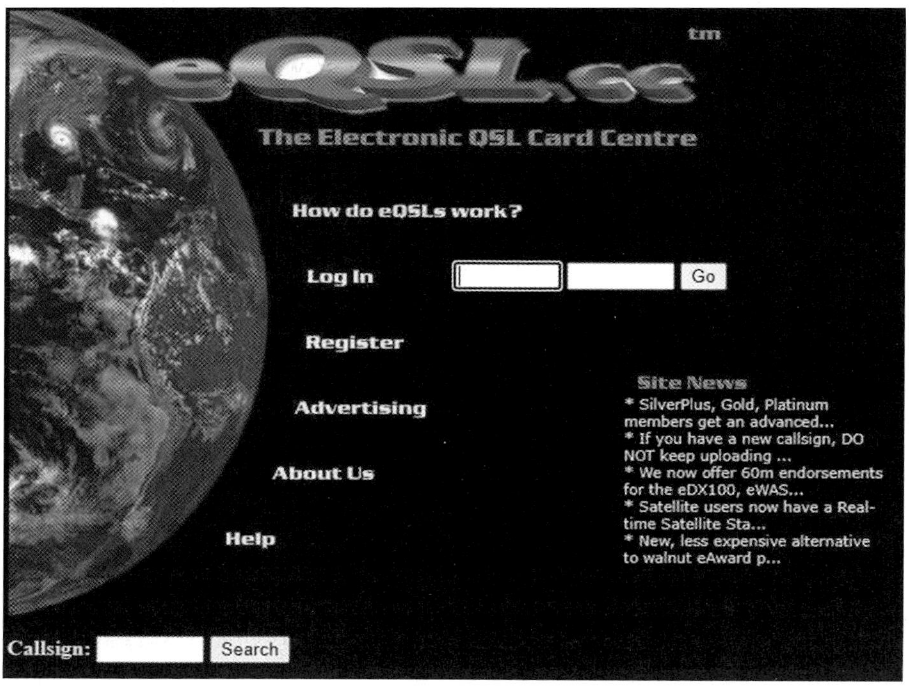

Home page of the Electronic QSL Card Centre – eQSL.

OQRS (ONLINE QSL REQUEST SERVICE)

Perhaps the major change in QSLing since the first *6 Metre Handbook* was published in 2008 has been the widespread acceptance of Online QSL Request Services (OQRS). This is a logical development for a number of reasons.

The basic premise is that you can order a direct or bureau card via an Internet site rather than have to mail a physical card with the associated cost, risk of it being lost *en route* and inevitable delays in the post and / or the bureau system. It also has the advantage that it is never the best idea to send cash (such as dollar bills) through the mail, while IRCs (International Reply Coupons) are being phased out by many postal administrations (signatories to the international postal conventions are still obliged to honour IRCs, but many no longer make them available for sale).

OQRS started with a number of DXpeditions making a request facility available on their website, usually in association with a capability to make a donation via PayPal (or other means) to cover mailing costs and perhaps help to cover other DXpedition costs. This move made a lot of sense – very few DXpeditions wish to retain incoming QSL cards so why collect them in the first place?

Many expeditions continue to offer OQRS via their own websites, but a powerful and user-friendly alternative has come to the fore in recent years by way of G7VJR's excellent Club Log **[16]**. Club Log offers not only expeditions but also individual DX stations and, indeed, any user the ability to receive QSL requests via the site. So, for example, if you don't wish to collect QSL cards for your more mundane day-to-day contacts, you can still receive and honour QSL requests from those who would like to have a card from you.

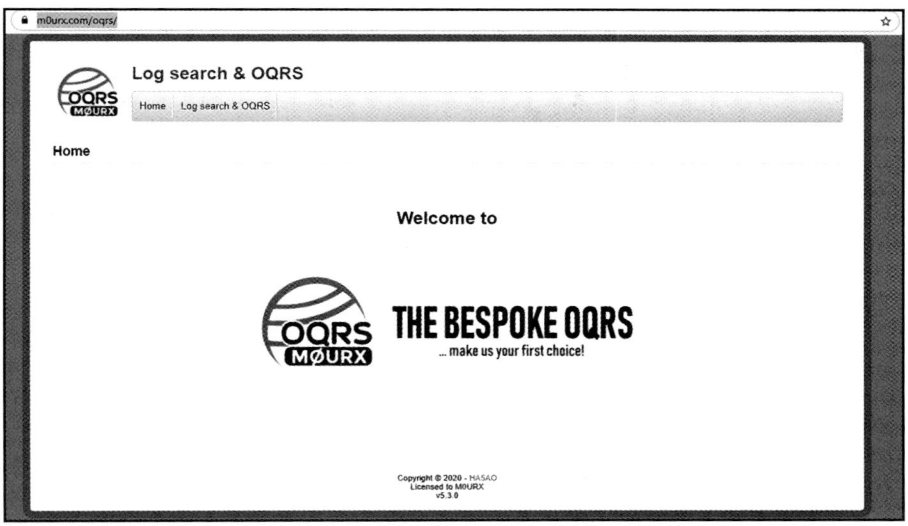

The OQRS page on the website of well-known UK QSL manager Tim Beaumont, M0URX.

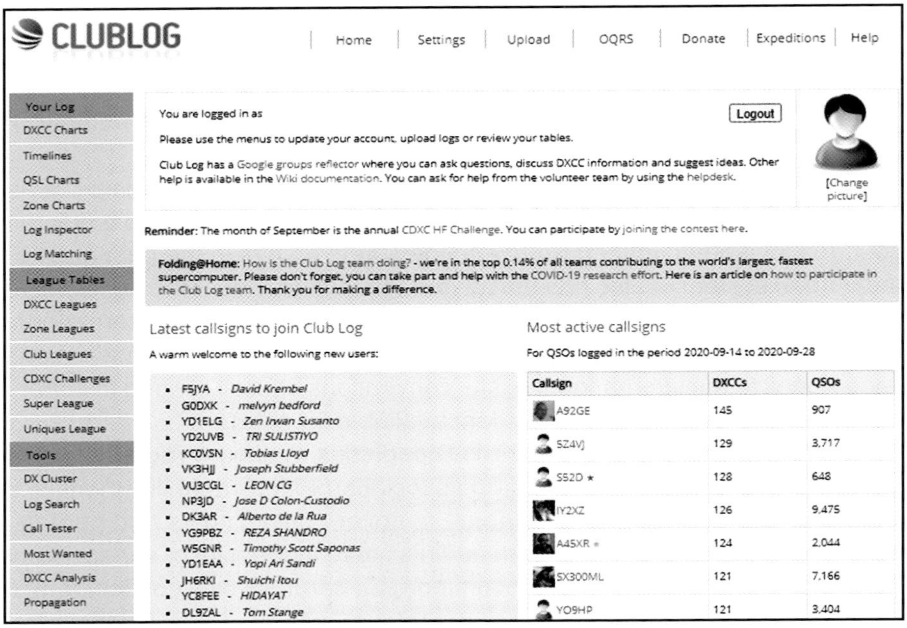

The home page of Club Log by G7VJR.

Club Log offers many other facilities too, as has been mentioned elsewhere in this book and new features are being added on a regular basis. It is well worth signing up and taking advantage of the facilities it offers.

REFERENCES

[1] *RSGB Yearbook*, published annually by the RSGB
[2] RSGB VHF Contests: www.rsgbcc.org/vhf
[3] CQ WW VHF Contest: www.cqww-vhf.com
[4] SMIRK: www.smirk.org
[5] ARRL Contest Calendar: www.arrl.org/contest-calendar
[6] OZ VHF Contests: www.qsl.net/oz6om/the50mhzopen/The50MHzOpen.html
[7] ZS VHF Contests: www.sarl.org.za/public/contests/contestrules.asp
[8] RSGB VHF Awards: https://rsgb.org/main/operating/amateur-radio-awards/vhfuhf-awards
[9] ZS6EZ list of 6m DXCC recipients: http://zs6ez.org.za/lists/dxcc-6m.txt
[10] ARRL list of DXCC holders and scores: www.arrl.org/dxcc-standings
[11] CQ awards: www.cq-amateur-radio.com/cq_awards/index_cq_awards.html
[12] UK Six Metre Group: www.uksmg.org
[13] IOTA: www.iota-world.org
[14] K1BV awards site: www.dxawards.com
[15] eQSL: www.eqsl.cc/qslcard/Index.cfm
[16] Club Log: www.clublog.org

Appendices

APPENDIX A – GLOSSARY OF TERMS AND ABBREVIATIONS

AM	Amplitude Modulation
ARRL	American Radio Relay League (the national body representing amateur radio in the USA)
CEPT	The European Conference of Post and Telecommunications Administrations
CW	Continuous Wave (refers to Morse transmissions)
DSP	Digital Signal Processing
DX	Long-distance (or rare) in the context of stations contacted
DXCC	DX Century Club (major international awards programme)
DXing	Chasing contacts with rare or distant stations (see DX)
DXpedition	An expedition specifically to activate a DX location
EMC	Electromagnetic Compatibility
EME	Earth-Moon-Earth, using the moon as a passive reflector to make long-distance contacts
ERP	Effective Radiated Power
FCC	Federal Communications Commission (US licensing body)
HF	High Frequency (officially 3 – 30MHz, but used by amateurs to refer to the bands 30 to 10m)
IARU	International Amateur Radio Union
IOTA	Islands on the Air (major awards programme)
ITU	International Telecommunication Union
LoTW	Logbook of The World
lpm	Letters per Minute
MS	Meteor Scatter
MUF	Maximum Usable Frequency
NCDXF	Northern California DX Foundation

NoV	Notice of Variation (to UK amateur licence)
OFCOM	Office of Communications (UK)
PC	Personal Computer
QSL Card	Card exchanged by radio amateurs to confirm a successful two-way contact
QSO	Used to refer to a contact between radio amateurs
RSGB	Radio Society of Great Britain
sked	Pre-arranged schedule
SMIRK	Six Metre International Radio Klub
SSB	Single Sideband
SWR	Standing Wave Ratio
TEP	Trans-Equatorial Propagation
UKSMG	UK Six Metre Group
UTC	Universal Coordinated Time (effectively the same as Greenwich Mean Time)
VHF	Very High Frequency (officially 30 – 300MHz, used by amateurs to refer to the 50, 70, 144 and, in North America, 220MHz bands)
WAS	Worked All States (an awards programme)
WAZ	Worked All Zones (an awards programme)
WSJT	Weak Signal by K1JT (a suite of programs developed for weak signal working)
Yagi	The most popular type of directional antenna (named after one of its Japanese inventors)

APPENDIX B – SOURCES OF INFORMATION

This Appendix lists a wide range of sources, including those referenced elsewhere in this book, where you can find information, software to download, and much else that is relevant to the subject matter of the book. Inevitably, the longest list is of websites, as these offer the most up-to-date sources and are easily accessible (and free!) However, it must be remembered that websites come and go, and their addresses change, but the list is believed to be accurate at press time.

GENERAL (HISTORY, CLUBS ETC)

Books

VHF / UHF Handbook, RSGB.
RSGB Yearbook, published annually by RSGB.
The ARRL Handbook, ARRL.
Radio Communication Handbook, RSGB.
The VHF / UHF DX Book, G3SEK and others, DIR Publishing, 1992 (republished by RSGB but now out of print).
The ARRL Operating Manual for Radio Amateurs, ARRL.
ARRL's VHF Digital Handbook, Steve Ford, WB8IMY, ARRL.
Guide to VHF/UHF Amateur Radio, Ian Poole, G3YWX, RSGB.

Articles

'What to Expect on 6', Bill Wageman, K5MAT, *QST* August 2004, pp49-51.
'Six Metre Transceivers', *Six News*, Issue 65, May 2000.

Websites

French bands history: http://oernst.org/hamradio/bande/bandesOM.html
Four metres website: www.70mhz.org
SMIRK: www.smirk.org
M5BXB: www.m5bxb.com
DK7ZB: www.qsl.net/dk7zb/
OE 6m Pages (lots of 6m links): www.qsl.net/oe4whg/6mlinks.htm
DX Zone: www.dxzone.com/catalog/Operating_Modes/50_Mhz
DK5YA VHF Page: www.vhfdx.de/6mtrs.html
UKSMG: www.uksmg.org
G3WOS: www.gare.co.uk/g3wos.htm
H44PT (G0BCG): www.h44pt.org.uk

EQUIPMENT & STATION DESIGN

Books

Twenty Five Years of Hart Reviews, Peter Hart, G3SJX, RSGB.
Amplifiers, Peter Rodmell, G3ZRS, published by author.
Computers in Amateur Radio (2nd edition), edited by Steve White, G3ZVW, RSGB 2013.
VHF / UHF Handbook, RSGB.
Remote Operating for Amateur Radio, Steve Ford, WB8IMY, ARRL 2010.

Articles

'There's a Remote Possibility...', David Gould, G3UEG, *RadCom* August 2005 p80, September 2005 p76, and October 2005 p76.

Websites

WinSDR: http://psn.quake.net/WinSDR
CW Skimmer: www.dxatlas.com/CwSkimmer
Fun Cube Dongle: www.funcubedongle.com
OZ2M transverter design: www.rudius.net/oz2m/70mhz/transverter.htm
G3WOS 6m amplifier design: www.gare.co.uk/amplifier/index.htm
Low Pass Filter: www.ham-radio.com/n6ca/50MHz/50appnotes/50tlpf.html
G4IFB / ZL2IFB on using Logger 32 and WSJT: https://tinyurl.com/y3akkgs7
JT Alert: https://hamapps.com
Grid Tracker: https://tagloomis.com/grid-tracker
DXLite: http://g7vjr.org/dxlite
'[Not] Living with 6m/50MHz interference!': www.gare.co.uk/noise/index.htm
Noise reduction video: https://youtu.be/qdQm17udCSo
ON4WW EMI – RFI page: www.on4ww.be/emi-rfi.html
4m Amplitec linear amplifier: www.amplitec.hu/ug_4_100_1000_gs31b_eng.html
FT-847 4m mods: http://oz1djj.geronne.dk/ft847_4m_mods.htm
Remote Control via Internet: www.dh7fb.de
RF Hamdesign: www.rfhamdesign.com
GM3SEK: www.ifwtech.co.uk/g3sek

ANTENNAS

Books

The ARRL Antenna Book, ARRL.
Antenna Compendium No 5, ARRL ('Two portable 6-Meter Antennas', designs for a 2-element quad and a 3-element Yagi).
ARRL's VHF / UHF Antenna Classics, ARRL.
RSGB Antenna File, compiled and edited by Steve Telenius-Lowe, 9M6DXX, RSGB, 2013.
Antenna Physics, Robert Zavrel Jr, W7SX, ARRL.
Magic Band Antennas for Ham Radio, Joel Hallas, W1ZR, ARRL.
Antennas for VHF and Above, Ian Poole, G3YWX, RSGB.
The ARRL Antenna Book, ARRL.

Websites

InnovAntennas: www.innovantennas.com
G0KSC: www.g0ksc.co.uk
YU7EF: www.yu7ef.com
G3WOS 6m antenna: https://www.gare.co.uk/6m_antenna
RF Hamdesign: www.rfhamdesign.com/products/powersplitters/2waysplitter/index.php
DK7ZB home page: www.qsl.net/dk7zb
GM3SEK on stacking: www.ifwtech.co.uk/g3sek/stacking/stacking2.htm
GM3WOJ on lattice towers: www.qsl.net/gm3woj/latticetower.htm
Messi & Paoloni: https://messi.it/en/home.htm
GM3SEK website: www.ifwtech.co.uk/g3sek
Cable loss data: www.qsl.net/dk3xt/cable.htm
U310 Preamp: www.ham-radio.com/n6ca/50MHz/50appnotes/U310.html

PROPAGATION

Books

Radio Auroras, Charlie Newton, G2FKZ, revised edition with new chapter by Neil Carr, G0JHC, RSGB 2012.
Sun, Earth & Radio, J A Ratcliffe, World University Library, 1970 (out of print).

Ionospheric Radio, Davies, ISBN 0 86341 186 X.
Radio Propagation Explained, Steve Nichols, G0KYA, RSGB.
Propagation and Radio Science, Eric Nichols, KL7AJ, ARRL.

Articles

'Sporadic E – A Mystery Solved?' Parts 1 and 2, *QST*, October and November 1997.
'Making the Most of Sporadic E at VHF', *RadCom*, January 2002, p34.
'Es Sporadic E propagation at VHF: a review of Progress and Prospects', Emil Pocock, W3EP, *QST*, April 1988.
'Meteor Scatter Communications', Clarke Greene, K1JX, *QST*, January 1986, pp14-17.
'Catch a Falling Star', Kirk Kleinschmidt, NT0Z, *QST*, October 1997, pp63-67.

Websites

JE1BMJ SSSP: https://tinyurl.com/y48vnukk
KH6/K6MIO 'Extreme Multi-Hop 50MHz Es': https://tinyurl.com/yxl8keho
QST TEP article, part 1: www.vhfdx.net/docs/qst_te_nov_1981_part1.pdf
QST TEP article, part 2: www.vhfdx.net/docs/qst_te_dec_1981_part2.pdf
TEP bibliography: www.dxmaps.com/tepbiblio.html
DF5AI: www.df5ai.net
Space Weather: www.spaceweather.com
'50MHz F2 Propagation Mechanisms': https://tinyurl.com/yx8ztruo
'A Seven Year Study of 50MHz Propagation': www.qsl.net/w/wa5iyx/cq/cq7208a.htm
Useful references and free software: www.sss-mag.com/prop.html#software
Solar Ham by VE3EN: www.solarham.net
Space Weather: www.spaceweather.com
K1SIX: www.k1six.com (with links to other good sites)
RSGB Propagation Studies Committee: www.rsgb.org.uk/psc
International Meteor Organization: www.imo.net
G3USF 50MHz Beacon List: www.keele.ac.uk/depts/por/50.htm
Meteor Scatter: www.dxzone.com/catalog/Operating_Modes/Meteors
'Investigation of the polarisation of 50MHz signals via Sporadic-E reflection', Chris Deacon, G4IFX, Ben Witvliet and Cathryn Mitchell: https://tinyurl.com/y5wonb36

OPERATING

Books

Amateur Radio Operating Manual, eighth edition, Mike Dennison, G3XDV, and Steve Telenius-Lowe, PJ4DX, RSGB, 2015.
ARRL Operating Manual, ARRL.

Websites

Daily DX: www.dailydx.com
OPDX Bulletin: www.papays.com/opdx.html
DX-World: www.dx-world.net
425 DX News: www.425dxn.org
DX Summit: www.dxsummit.fi
ON4KST: www.on4kst.com/chat/start.php
DXLite: http://dxlite.g7vjr.org
CW Skimmer: www.dxatlas.com/CwSkimmer
RBN: www.reversebeacon.net
UK 4m reflector: http://groups.yahoo.com/group/fourmetres
DX Code of Conduct: http://dx-code.org

G3WOS list of UK 'firsts': www.gare.co.uk/ukalltime.htm
List of 6m allocations: www.secornwall.pwp.blueyonder.co.uk/licensing.pdf
WinSDR: http://psn.quake.net/WinSDR
CW Skimmer: www.dxatlas.com/CwSkimmer

WEAK SIGNAL WORKING

Books

Getting Started in EME, Stephen Appleyard, G3PND, and Philip Malme, G4PQP, RSGB.
FT8 & PSK, Rob Walker, G3ZJQ, RSGB.
Work the World with JT65 & JT9, Steve Ford, WB8IMY, ARRL.
Digital Modes for all Occasions, Murray Greenman, ZL1BPU, RSGB.
Elimination of Electrical Noise, Don Pinnock, G3HVA, RSGB.

Articles

'I Don't use Digital, Honestly I Don't', Ken Osborne, G4IGO, *SixNews*, Issue 94 (Feb 2008).
'EME with JT65', Gene Zimmerman, W3ZZ, *QST*, June 2005, pp80-82.
'Work the World with WSJT-X, Part 1: Operating Capabilities', *QST* October 2017, pp30-36 (a general introduction to the digital modes supported by WSJT-X).
'Work the World with WSJT-X, Part 2: Codes, Modes, and Cooperative Software Development', *QST* November 2017, pp34-39 (dips further into mathematics, Forward Error Correction etc).
'The FT4 and FT8 Communication Protocols', Steve Franke, K9AN, Bill Sommerville, G4WJS, and Joe Taylor, K1JT, *QEX* July / August 2020.
'The MSK144 Protocol for Meteor-Scatter Communication', Steve Franke, K9AN, and Joe Taylor, K1JT, *QEX* July / August 2017.

Websites

WSJT page: http://physics.princeton.edu/pulsar/K1JT
ZL2IFB FT8 Guide: www.g4ifb.com/FT8_Hinson_tips_for_HF_DXers.pdf
Dimension 4: www.thinkman.com/dimension4/download.htm
ON4KST: www.on4kst.com/chat/start.php
Ping Jockey: www.pingjockey.net/cgi-bin/pingtalk
International Meteor Organization: www.imo.net
Virgo Meteor Sky View (DL1DBC): www.dl1dbc.net/Meteorscatter
Wikipedia on EME: http://en.wikipedia.org/wiki/EME_(communications)
W7GJ: www.bigskyspaces.com/w7gj
InnovAntennas: http://innovantennas.com
G3WOS antennas: www.gare.co.uk/6m_antenna
GM4JJJ MoonSked: www.gm4jjj.co.uk/MoonSked/moonsked.htm
JT4/JT65 Terrestrial Link (chat page): www.chris.org/cgi-bin/jt65talk
Wikipedia on Faraday rotation: http://en.wikipedia.org/wiki/Faraday_rotation
IW5DHN: www.qsl.net/iw5dhn/when.htm
N1BUG: www.g1ogy.com/www.n1bug.net/prop/eme.html
DF9CY: www.df9cy.de/serendipity/index.php
PSK31 'Official' Homepage: http://bipt106.bi.ehu.es/psk31.html
PSK63: www.qsl.net/kh6ty/psk63
JS8Call: http://js8call.com
G0CHE website: www.g0che.co.uk
OH5IY: www.kolumbus.fi/oh5iy
VK3UM EME software: www.vk3um.com
SM2CEW EME: www.sm2cew.com

6m AND 4m AROUND THE WORLD

Websites

JI1CQA: http://ji1cqa.sakura.ne.jp/index.shtml
JA1RJU (history, in English): https://tinyurl.com/y2ru5cm6
ZS6EZ: http://zs6ez.org.za/lists/50zs1st.txt
EY8MM: www.ey8mm.com/magic-band-50-mhz
PY5CC: www.50mhz.com/py.html
VK3OT tape archives: www.telesupport.nl/wave/vk3ot/index.htm

PORTABLE AND EXPEDITION OPERATING

Books

RSGB Yearbook, published annually by RSGB.
World Licensing and Operating Directory, by Steve Telenius-Lowe, 9M6DXX, RSGB, 2008.

Websites

'Considerations for Successful 6m EME DXpeditions', Lance Collister, W7GJ; https://tinyurl.com/p7lpz9o
Club Log: www.clublog.org
G3WOS 6m amplifier: www.gare.co.uk/amplifier
Cracao 6m EME DXpedition: https://tinyurl.com/y2u26mwa
UK Six Metre Group: www.uksmg.org
NCDXF: www.ncdxf.org

CONTESTS, AWARDS, QSLING

Books

Amateur Radio Contesting for Beginners, Doug Grant, K1DG, ARRL.

Websites

RSGB VHF Contests: www.rsgbcc.org/vhf
CQ WW VHF Contest: www.cqww-vhf.com
SMIRK: www.smirk.org
ARRL Contest Calendar: www.arrl.org/contest-calendar
OZ VHF Contests: www.qsl.net/oz6om/the50mhzopen/The50MHzOpen.html
ZS VHF Contests: www.sarl.org.za/public/contests/contestrules.asp
ZS6EZ list of 6m DXCC recipients: http://zs6ez.org.za/lists/dxcc-6m.txt
ARRL list of DXCC holders and scores: www.arrl.org/dxcc-standings
CQ awards: www.cq-amateur-radio.com/cq_awards/index_cq_awards.html
IOTA: www. www.iota-world.org
RSGB VHF Awards: https://rsgb.org/main/operating/amateur-radio-awards/vhfuhf-awards
K1BV awards site: www.dxawards.com
eQSL: www.eqsl.cc/qslcard/Index.cfm
Club Log: www.clublog.org

APPENDIX C – UK 6M VOICE REPEATERS

Callsign	6m Channel	Output freq	Input freq	Loc	Location	CTCSS tone (Hz)	Keeper	Status (see Notes)
GB3AE	R50-01	50.7200	51.2200	IO71MQ	PEMBROKE DOCK	94.8	GW4OZU	1
GB3AM	R50-13	50.8400	51.3400	IO91QP	AMERSHAM	77	G0WTZ	1
GB3DB	R50-06	50.7700	51.2700	JO01HR	DANBURY	110.9	G6JYB	1
GB3EC	R50-12	50.8300	51.3300	JO02HE	BURY ST EDMUNDS	110.9	G1YFF	1
GB3EF	R50-01	50.7200	51.2200	JO02NF	STOWMARKET	110.9	G1YFF	1
GB3FH	R50-06	50.7700	51.2700	IO81OH	SOMERSET	77	G4RKY	1
GB3FX	R50-10	50.8100	51.3100	IO91OF	FARNHAM	82.5	G4EPX	1
GB3GC	R50-02	50.7300	51.2300	IO70WN	GUNNISLAKE	77.0	M0WMB	3
GB3HM	R50-11	50.8200	51.3200	IO93GA	BELPER	71.9	G8IQP	1
GB3HX	R50-05	50.7600	51.2600	JO00HV	HASTINGS	103.5	G8PUO	1
GB3LP	R50-04	50.7500	51.2500	IO83MK	LIVERPOOL	77	M1SWB	1
GB3PX	R50-07	50.7800	51.2800	IO92XA	ROYSTON	77	M0ZPU	2
GB3SL	R50-02	50.7300	51.2300	IO75XX	KILSYTH	103.5	GM4COX	1
GB3SX	R50-08	50.7900	51.2900	IO93BA	STOKE ON TRENT	103.5	G4SCY	1
GB3TY	R50-07	50.7800	51.2800	IO74	CARRICKFERGUS	110.9	GI6DKQ	1
GB3UM	R50-03	50.7400	51.2400	IO92IQ	MARKFIELD	77	M1NAS	1
GB3VI	R50-15	50.8600	51.3600	IO92BJ	SUTTON COLDFIELD	67.0	G8NDT	4
GB3WX	X-Band	50.5200	29.2100	IO81VC	MERE WILTSHIRE	77	G3ZXX	4
GB3WY	R50-09	50.8000	51.3000	IO93EP	WAKEFIELD	82.5	G1XCC	1
GB3XS	R50-14	50.8500	51.3500	IO92MG	NORTHAMPTON	77.0	G7SYT	2
GB3ZW	R50-10	50.8100	51.3100	IO82HL	NEWTOWN POWYS	103.5	GW4IQP	4
GB3ZY	R50-09	50.8000	51.3000	IO81QJ	BRISTOL	77	G4RKY	1

Notes: 1 Operational, 2 Not operational, 3 Reduced output, 4 Licensed.

APPENDIX D – 6M AND 4M SIMPLEX GATEWAYS

SIMPLEX GATEWAYS 6M

Call	Keeper	Code	Freq MHz	Modes	Loc	Region	Location	Status (see Notes)
MB7IAR	2M0SBP	88.5Hz	50.5300	A	IO76BV	SCO	ARISAIG	1
MB7IBR	G0WYG	103.5Hz	50.5200	A	JO01AK	SE	BROMLEY	1
MB7INK	G4YNK	82.5Hz	50.5200	A	IO83SO	NOR	BOLTON	1
MB7ITW	M0TWS	71.9Hz	50.5300	A	IO93GJ	MID	SHEFFIELD	1
MB7PA	G7AJP	71.9Hz	50.5200	A	IO93WH	MID	LOUTH	1

SIMPLEX GATEWAYS 4M

Call	Keeper	Code	Freq MHz	Modes	Loc	Region	Location	Status (see Notes)
MB7AAR	2M0SBP	88.5Hz	70.3625	A	IO76BV	SCO	ARISAIG	1
MB7ABR	G0WYG	103.5Hz	70.4125	A	JO01AK	SE	BROMLEY, KENT	1
MB7ACB	MW0VTK	77Hz	70.3875	A	IO72WS	WM	Nr BARMOUTH	1
MB7AKE	GM7SVK	103.5Hz	70.4125	A	IO75	SCO	GLASGOW	1
MB7AM	G0WTZ		70.2600	A	IO91QP	SW	Nr AMERSHAM	1
MB7ANM	M0NFI	82.5Hz	70.3875	A	IO83TR	NOR	OSWALDTWISTLE	1
MB7ARH	G0SPH	82.5Hz	70.3625	A	IO83RE	NOR	WINSFORD	1
MB7ARX	G4IRX	77Hz	70.3500	A	IO92	MID	NOTTINGHAM	4
MB7FM	G0WTZ	77Hz	70.4375	A	IO91PS	SW	Nr TRING	1

Notes: 1 Operational, 2 Not operational, 3 Reduced output, 4 Licensed.

APPENDIX E – 4M BEACONS

Freq	Callsign	Location	Loc	Contact	Pwr	Antenna	Notes
70.000	GB3BUX	Buxton	IO93BF	G4IHO	20	2x turnstile	
70.005	V51VHF		JG82IE		35	6 el N	
70.005	WG2XPN	Bedford, VA	FM07FM	WA1ZMS		3 el 60°	CW, 3kW ERP
70.007	GB3WSX	Yeovil	IO80QW	G3ZXX	15	5 el 70°	F1A
70.008	OH1		KP11	OH6DD			Planned
70.010	J5FOUR/B	Bissau	IK21EV	J5JUA	20	Dipole	CW
70.010	ZS6JON	Krugersdorp	KG33VV	ZS6JON	20	4 el 0°	
70.012	OX4MB	Kangerlussuaq	GP47TA	OZ1DJJ	10	GP	
70.014	S55ZRS	Mt. Kum	JN76MC				Planned
70.016	GB3BAA	Tring, Herts	IO91PS	G0RDI	20	Dipole	
70.016	SV5FOUR	Rhodes	KM46CK	SV5BYR	5	5 el 310°	
70.018	OH2FOUR	Lohja	KP20DH	OH2LIY	30	Vert dipole	
70.020	GB3ANG	Dundee	IO86MN	GM4ZUK	100	3 el 160°	
70.021	OZ7IGY	Jystrup	JO55WM	OZ7IS	25	Big wheel	PI4+CW+carrier
70.023	9A0BFH		JN85JO		10	Dipole	
70.023	PI7EPO	Rijswijk	JO22EA	PA5DD	20	Halo	F1A, planned (1)
70.024	LZ2HV/B	Veliko Tarnovo	KN23TB	LZ2HV	20	Dipole	A1A
70.025	GB3MCB	St Austell	IO70OJ	M0PHM	40	2 el 45°	
70.025	ZS6WAB	Pietersburg	KG46RC		25	2x 9 el 0°	
70.026	SR9FHK	Kochanow	JO90VC	SP9SVH	10	GP	A1
70.027	GB3CFG	Carrickfergus	IO74CR	GI0GDP	20	3 el 45°/135°	PI4+CW+carrier
70.029	S55ZMB		JN76VK	S51DI	6	4 el 310°	A1A+15s keydown
70.030	HG7BVC						
70.031	G4JNT/P	Blandford	IO80UU	G4JNT	0,6	Dipole	(2)
70.033	OH5RBG	Kouvola	KP30HW	OH5IY	5	3 el 232°	A1A
70.035	OY6BEC	Tórshavn	IP62	OY9JD	25	2 el 135°	F1A
70.037	ES1VHF	Tallinn	KO29IK	ES1NI	10	Dipole SW/NE	F1A+10s carrier
70.040	SV1FOUR	Athens	KM18UA	RAAG	20	5 el NW/S	(3)
70.043	PA1O	Helmond	JO21TL	PA1O	4	Mag loop	F1A 250 Hz
70.045	OE5QL	Linz	JN78CJ	OE5MPL	1/5	Big wheel	(4)
70.047	OX4M	Ittoqqortoormiit	HQ90AL	OZ1DJJ	25	Dipole N/S	PI4+CW+carrier
70.048	4O0BCG	Podgorica	JN92PK	4O7JAZ	10	X-dipoles	PI4+CW+carrier
70.050	GB3RAL	Oxfordshire	IO91IN	G3VPW	10		(5)
70.055	SV9FOUR	Crete	KM25EH	SV9GPV	7.5	3 el	
70.057	TF1VHF	Álftanes	HP84WL	TF3ML	20	Dipole NW/SE	
70.060	HG1BVC/B	Horman	JN87FI	HA1YA	10	CX-dipoles	

Freq	Callsign	Location	Loc	Contact	Pwr	Antenna	Notes
70.063	LA2VHF	Trondheim	JP53EG	LA1K	25	4 el 15°	A1A+10s carrier
70.065	LA5VHF	Kristiansand	JO48AD	LA4YGA	20	GP	F1A
70.071	(OZ7IGY)						Reserved
70.073	SV8FOUR/B		KM08HQ				
70.075	LA4VHF	Hellvik	JO28WL	LA3EQ	15	Dipole N/S	FSK 330Hz
70.075	Z35MSK		KN01RX		10	Omni	
70.081	LA7VHF	Bardu	JP99EC	LA5TFA	20	4 el 180°	
70.081	HG8BVC/B	Kisrate	KN06HT	HA8MV	10	X-dipoles	
70.085	A92C/B	Mina Salman	LL56HE	A92GE	10	Vert dipole	F1A
70.087	SV2JAO/B		KN10CL	SV2JAO	10	4 el	
70.088	I0JX/B	Rome	JN61HV	I0JX	6	4 el 340°	
70.090	HG9BVC	Kis-kohat	KN08FB	HA9MDP	10	2 el 240°	
70.090	IS0GRB/B	Cagliari	JM49OF	IS0GRB	10	4 el N/NE	A1A
70.090	UN3M/B	Aksai	LO61MD	UN3M	10	4 el 275°	A1A
70.092	IW9GDC/B	Messina	JM78SD	IW9GDC	12	Big wheel	PI4+CW+carrier
70.10025	SR3FHK	Konin	JO92AD	SP3RNZ		Turnstile	F1A
70.102	SR8FHJ		KN09RR	Sp8WJW	3	X-dipoles	
70.107	IZ1DYE/B		JN35PA	IZ1DYE	7	Dipole N/S	
70.110	SR5FHR	Kampinos	KO02GH	SP5XMU	5	Vertical	A1A planned
70.111	IQ5PT/B		JN53JV	IK5PWB	4	Halo	A1A
70.11355	B4CY	Cyprus	KM64FT	5B8AV	10	6 el 315°	
70.117	OK0EE	Strá•ek	JN89CK	OK1CDJ	10	Dipole E/W	
70.130	EI4RF	Dublin	IO63VE	EI9GK	25	5 el 45°/135°	(6)
70.159	CS5BALG		IM67AH				
70.160	CS4BFM		HM76KR	URA	10	Dipole	F1A 150Hz
70.161	LX0FOUR	Bourscheid	JN39AV	LX1JX	10	Dipole	
70.163	CS3BFM	Pico do Silva	IM12OR	CS3MAD	10	4 el to Eur.	A1A
70.164	CS5BFM	Santarem	IM59QD	CT1ARR	10	Dipole	A1A
70.165	CU8B		HM49KL		20		F1A
70.220	D44TD/B		HK86NO	D44TD	5	4 el to Eur.	F1A
70.375	EI4SMR	40km NW Cork	IO52MA	EI7GL	10	Vertical	FM 45s (7)
70.3875	MB7NS	Banstead	IO91VH	G0OLX	10	Dipole	FM 120s (7, 8)
70.4375	MB7FM	Chiltern Hills	IO91PS		10	Dipole	FM (7)
75.300	ZD8DUB	Ascension	II22TB		8	4 el 340°	

Notes: (1) Not yet licensed, (2) Slow FSK data +85Hz shift and CW, (3) NW May-Aug, S Sep-Apr, (4) HH:00-HH:02, HH:15-HH:17, HH:30-HH:32, HH:45-HH:47, (5) Odd minutes CW + carrier, Even minutes JT65B, (6) Yagis sequenced, (7) Simplexer, (8) Off air due to relocation.

APPENDIX F – 6M BEACONS

Freq	Call	Town	Loc	Pwr	Mode	Last Rept
40050	GB3RAL	Nr Didcot	IO91IN		A1/JT65B	NonOp
40071	OZ7IGY	Jystrup	JO55WM	20	A1/P14	0518
50000	GB3BUX	Nr Buxton	IO93BF	25	A1	0220
50000.5	YO7AQF	Pitesti	KN24KV	1		0519
50000.2	F1ZGD	Toulon	JN23XD	2		0320
50001	BV2B	Taipei	PL05SA	10	A1	0617
50001	VE1UW	Pictou NS	FN85QN		A1	0819
50001	VE9SIX	Keswick Ridge	FN21TR		A1	NonOp
50003	VO1FRR	Nr St John NL	GN37JS	10	A1	IRREG
50004	A47RB	Oman	LL93FO			0518
50004	I0JX	Rome	JN61GW	10	CW	1017
50004.7	I0JX				FT8	0919
50004.5	AH2G	Guam	QK23	50	A1	0718
50005	OZ4BMS		JO75KB		CW/MGM	0619
	OZ4BHM?					0619
50005	EI0SIX	Enniskerry	IO63VE	30	PI4/CW	0719
50006	GB3MCB	St Austell	IO70OJ	40	Sync CW/PI4	0120
50006	GB3NGI	Ballycastle	IO65VB		Sync CW/PI4	0519
50006	A71A	Doha	LL55SH	10	A1	0819
50006	IW9GDC	Messina	JM78SD		P14+CW	0719
50007	EA7URC	Cordoba	IM77RQ		A1	0519
50007	HG1BVB	Horman	JN87FI	20	F1	0220
50007.8	DU1EV	Metro-Manila	PK04MP	10	A1	0719
50008	K0GUV	Park Rapids, MN	EN26	8	A1	1119
50008	HI8W	Ft Resolue	FK48	5	A1	0319
50008	CE1BB					0319
50008.6	J88ARC	St Vincent	FK93		A1	IRREG
50008.8	VA2ZFN	Mt Kanasuta, QC	FN08HE	4	A1	0919
50009	4S7B	Colombo	MJ96XV	25	A1	0419
50009.5	VE3WCC	Nr Ottawa	FN15WC	1	A1	0719
50010	LU7FTF	Castelar SF	FF88XK	5	A1	0220
50010	SV9SIX	Iraklio	KM25NH	30	F1	0818
50010	JA2IGY	Mie	PM84JK	10	A1	0519
50010.5	K8MMM	Novelty, OH	EN91	90	A1	0220
50011.2	ZL1SIX	Bay of Islands	RF64VS	25	A1	1119
50011.9	OX3SIX	Tasiilaq	HP15EO	100	?	0819
50012	OH1SIX	Ikaalinen	KP11QU		A1	NonOp
50012	V73SIX	Roi Namur	RJ39RJ	40	A1	0519
50013	BG6CJR	Ningguo	OM90LO			0619
50015.9	XE2K	Mexicali	DM22FP	20	A1	1219
50016	GB3BAA	Nr Tring	IO91PS	10	A1	1019

Freq	Call	Town	Loc	Pwr	Mode	Last Rept
50016	HP1AFF		FJ09GB	5	A1	0419
50017	JA6YBR	Miyazaki	PM51RT		A1	0819
50017	VO1STA	St Anthony, NL	GO21FJ		A1	IRREG
50018.7	VE4ARM	Austin, MB	EN09MW	50	A1	1119
50018	PR8ZIX	Imperatriz MA	GI64GL	45	A1	1218
50018	XE1RCS	Cerro Gordo	EK09OS	50	A1	IRREG
50018	5B4CY	Mandria	KM64KU	15	F1	0919
50019.7	4O0BMN	Podgonca	JN92PK	20	A1	0918
50020	YV6CR	El Tigre	FJ78VV		A1	0319
50020	VE8WD	Yellowknife, NT	DP22	20	A1	?
50020.1	ER1SIX	Nr Kishinev	KN47JG	1	A1	0919
50021	CX1CCC	Montevideo	GF15VD	5	A1	0817
50021.8	LX0SIX	Bourscheid	JN39AV	5	A1	0120
50021.4	UN1SIX		MN83KE	12		1118
50022	F6KBO	Tregueux	IN88PL		A1	0119
50022	YS1AG	San Salvador	EK53IP	40	A1	0220
50023	VE9BEA	Fredericton, NB	FN65OX		A1	0117
50023	SR5FHX	Warszawa	KO02KH	3	A1	0617
50024.0	ZL2WHO	Waipuna Ridge	RF70OM	200	A1	0220
50025	9H1SIX	Attard, Malta	JM75FV	7	F1	1019
50025	YV4AB	Valencia	FK60AD	15	A1	0220
50026	PP8AA	Manaus	GI06		A1	0918
50027	JE7YNQ	Fukushima	QM07	50	A1	0719
50027	CN8MC	Rabat	IM64OF	3	A1	0817
50027	VO1GAM	Lewisporte, NL	GN29LF	10	A1	0918
50027.7	SR3FHB		JO91CQ			NonOp
50028	ZP5WYC					1017
50029.3	9A0BHH		JN85JO			0120
50029.7	YM4SIX					1019
50030	VY1DX	Whitehorse, YU	CP20KR	20	A1	0219
50030	ZL2MHB	Napier	RF80KL	10	A1	0119
50031.5	CS5BCP		IM59QM			0517
50031.5	YU1EO		KN04ML			0817
50032V	JR0YEE	Niigata	PM97	2	A1	0517
50032	PT9AA	?				0220
50032.9	ZD8VHF	Ascension I	II22TB	50		NonOp
50033	VE7FG	Prince George, BC	CO83	50	A1	0619
50033	VE2RCS	Lachute, QC	FN25		A1	0819
50034	VA5MG	Nr Little Bear Lake	DO74PH	15	A1	0811
50033	OH5SHF	Kouvola	KP30HV	20	F1	QRT/QSY
50034.2	CS3BSM	Santo de Sierra	IM12OR	40	A1	0419
50035	IZ3OHR		JN55LJ			0716
50035	VO1SEP	Placentia, NL	GN37AE		A1	IRREG
50035	US0SU / US0SIX		KN28JW			IRREG
50036	VE4VHF	Headingly, MB	EN19	35	A1	1219

Freq	Call	Town	Loc	Pwr	Mode	Last Rept
50036	OA4B	Lima	FH17	13	A1	1019
50036V	CS5BALG	Sitio dos Cavalos	IM67AH	40	A1	0819
50036.5	SR8FHL	Lublin	KO11HF	4	A1	0120
50037	JR6YAG	Okinawa	PL36		A1	0719
50037.7	IK7LMX	Brindisi	JN80XP		A1	IRREG
50037	J68HZ		FK83MM		A1	0916
50038	VY0YHK	Gjoa Haven	EP28BP	15	A1	0919
50038	PY1ESP				A1	0120
50038.5	FR1GZ	Reunion I	LG79RG	10		0716
50039	VO1ZA	St Johns, NL	GN37	10	F1	0819
50039	FY7THF	Kourou	GJ35	10		0917
50040	C6AFP	Abaco			A1	0219
50040	ZL3SIX	Christchurch	RE66DL	50	A1	0120
50040	SV1SIX	Nr Athens	KM17UX		F1	0120
50040	VA2YKT	Rapide Blanc, QC	FN37LS		A1	0719
50041.1	CP6B	Santa Cruz da la Sierra	FH82	40	A1	0318
50041.2	LZ0SIX		KN32ER			0818
50041.7	ED6YAI	Menacor	JM19NL	5	A1	0819
50043	6K2EJJ	Namyangju	PM37NP	30	A1	0819
50043	9Y4D	FK90				0619
50043.2	ZL1VHF	Auckland	RF73MB	15	A1	1216
50044	ZS6TWB	Pdokwane	KG46RC	30	A1	0519
50044.7	YU1KM					0617
50045	OX3VHF	Qaqortoq	GP60QQ	20	F1	0719
50045	SR2FHM	Gdansk	JO94II	7	A1A	NonOp
50045.4	OK1DX		JN69SD			0518
50045.73	VK8RAS	Alice Springs	PG66WH	20	A1	0220
50046.2	PY2VA		GG66KU		A1	0219
50047	PY4AQA	Barbacena	GG88DS	30	A1	0119
50047	ZF1EJ	Cayman Is	EK99IG		A1	0617
50047	OX6M		HQ90AL	25	PI4+CW	0819
50047	YU0ZNI	SSE Beograd	KN03WH	1	F1	0120
50047.5	A62		LL75SJ	?		0320
50047.6	LU5FF		FF99RF		A1	IRREG
50048.8	LU1WFU		FE64AC			0220
50048.8	VY0SNO	Iqaluit, NU	FP53RR	30	A1	0819
50049	CE8B	Nr Punta Arenas	FD46NX	50	A1	0218
50049.6	LZ0SJB	Sliven	KN32DR	2		0120
50050	ZS6JON	Krugersdorp	KG33VV	20	A1	0416
50050	GB3RAL	Nr Didcot	IO91IN		A1/JT65B	
50050	TG9ANF	Guatemala City	EK44SM	19	A1	IRREG
50050.6	E51USA	Rarotonga	BG08CT	20	A1	0516
50050.7	LU8DCH	Martin Garcia I.	GF05VT	4	A1	0320
50051	6L0NJ	Gwangju	PM35KE	3	A1	0719
50052.7	V47JA	St Kitts	FK87PG	2	A1	1019

Freq	Call	Town	Loc	Pwr	Mode	Last Rept
50053	VE1PZ	Nr Pringle Lake, NS	FN95AI	25	A1	0819
50054	OZ6VHF	Ribe	JO57EI	25	A1	1219
50054.3	LZ0MON	Montana	KN13OJ			0120
50055	ZL3MHB	Westland	RE57MI	20		12185
50055.6	N9JXY	Auburn, IN	EN71LI	1	A1	0617
50056	VA3BYA	Mississauga, ON	FN03DM	3	A1	0617
50056	VE3TDG	Alfred, ON	FN25MM	8	A1	0919
50057	VK7RAE	Don Hill	QE38DU	20	A1	0120
50057	W2DLL	Buffalo, NY	FN02OU		A1	0120
50057	LU2AW	Buenos Aires				0119
50057	IT9X	Messina	JM78LB		A1/JT65B	0519
50057	TF1SIX	Fludir	HP93QU	8	A1	0819
50057.5	VK4RGG	Nerang	QG62QA	6	A1	0419
50058	HB9SIX	Alp Scheidogg	JN47LH	7	A1	0919
50058	OE3XLB	Maiersdorf	JN87AT	10		0719
50058.5	VE3UBL	Ajax, ON	FN03	10	A1	0220
50058.5	W4CBX	Bristol, TN	EM86		A1	0120
50059.91	4TQR	Birmingham, AL	EM63OM	4	A1	0220
50060	SK6QW	Nr Mariestad	JO68WR	8	A1	0067
50060	K5AB	Georgetown, TX	EM01DP	20	A1	0619
50060.0	GB3RMK	Nr Inverness	IO77UO	40	F1	0220
50060.0	WB5LLI	Kenner, LA	EM40	20	A1	0918
50060	HK6FRC	Neiva	FJ24DM		A1	IRREG
50060	WB0RMO	Fairbury, NE	EN10	25	A124	1219
50060	LU7HCS	Cordoba	FF78	A1		IRREG
50060	K4KNS	Hephzibah, GA	EM83	A1	24	0120
50060	N1NSP	Newport News, VA	FM17RD	A1	24	1119
50060.5	K7CA	Las Vegas, NV	DM26KC	50	A1	0220
50060.55	K9APW	Mt Toreb, WI	EN53DA	10	A1	0120
50061	N4RT	Bromley, AL	EM60BT	5	A1	0218
50061	N5SYV	Shreveport, LA	EM32	5	A1	0120
50061	WG2Z	Andover, NJ	FN21OA	100	A1	0618
50061	W3HH	Nr Ocala, FL	EL89	3	A1	1119
50061	KD4AOZ	Watkinsville, GA	EM83		A1	0819
50061.5	KZ9O	Altoona, WI	EN46	30	A1	0120
50061.9	K4PAR	Eatonton, GA	EM83II		A1	0519
50062	W7KNT	Nr Hall, MT	DN36HO	A1		0919
50062.0	W3PIE	Uniontown, PA	FM09DW	A1		1019
50062.5	GB3NGI	Ballycastle	IO65VB	30		0819
50062.5	W9JN	Carson, WI	EN54DM	20	A1	0120
50062.7	LZ0SIX		KN32ER	1		0819
50062.7	K7EMX		DN30XP	1	A1	0220
50062.6	W9RAS	Cassopolis, MI	EN71	1	A1	0120
50063	W8RS		EN71AV			1119
50062.9	K1MS	Littleton, MA	FN42FM	10	A1	1219

Freq	Call	Town	Loc	Pwr	Mode	Last Rept
50062.9	LU2ERC	Ensenada, BA	GF15AD		A1	0220
50063	LY0SIX	Vilnius	KO24PS	7	A1	0819
50063	WR7NV	Las Vegas, NV	DM25GV		A1	1218
50063	KB6BKN	Novato, CA	CM88RC		A1	0719
50063.4	N0SAP	Nixa, MO	EM37IB	8	A1	0719
50064.0	W3APL	Laurel, MD	FM19NE	7	A1	0120
50064	WB7PMP	Cleveland, NC	EM95PU	20	A1	0320
50064.2	KH6HI	Ewa, HI	BL01	50	A1	0119
50064.5	N1VVV	Windham, ME	FN43SS	25	A1	0717
50064.8	K0IBM	Denver, CO	DM79NQ	25	A1	0120
50065	GB3IOJ	Jersey	IN89WE		A1	NonOp
50065.0	KA0CDN	Aurora, CO	DM79OS	50	A1	1219
50065	N1CDN	Seminole, FL	EL87OU		A1	0919
50065	KE5JXC	Pecan Island, LA	EL39SP	5	A1	1119
50065	K8TB	Holland, MI	EN62WU	5	A1	0120
50065.9	W5GPM	Bartlesville, OK	EM26		A1	0120
50066	WA1OJB	Bowdoin, ME	FN54AA	30	A1	1219
50066	OE3XAC	Kaiserkogel	JN78SB	10	A1	0120
50066.2	W5SIX	Horse Mountain,NM	DM54WA	0.5	A1	0220
50067	WZ8D	Loveland, OH	EM89BI	100	A1	0220
50067	W4FWS	Orlando, FL	EL98HN		A1	0220
50067	OH9SIX	Pirttikoski	KP36OI	35	A1	QSY
50067.5	N8PUM	Champion, MI	EN66AL	50	A1	0120
50067.6	K5KDX	Van Buren, AR	EM25SM		A1	0919
50067.7	XE2O	Monterey	EL05		A1	0819
50067.7	K0EC	Louisville, CO	DM69EM		A1	1119
50067.8	LZ1WF	Plovdiv	KN22HI		A1	0819
50068	W1XQ	Sugar Land, TX	EL29	10	A1	0518
50068	9K2GS					1119
50068	N4LR	Fairburn, GA	EM73AL		A1	0220
50068.5	N2MH	West Orange, NJ	FN20US		A1	0220
50068.6	W4TTU	Cody, WY	DN54LM		A1	0619
50068.9	K6FV	Woodside, CA	CM87UL	100		0719
50069	N8GTX	St Clair Shores. MI	EN82	1	A1	NonOp
50069	WA3TTS	Pittsburgh, PA	EN90XL	1	A1	0719
50069	FM1ZAC	Martinique	FK94NL	15	A1	Op?
50069	W9RCA	Indianapolis, IN	EM69US	14	A1	1119
50069	N4BRF		EL96VK		A1	0818
50069.5	YO5PBG	Baia Mare	KN17SP	3	A1	0120
50069.2	YO4KRB	Constanta	KN43HE		A1	1118
50069.2	YO4FYQ		KN44FD	2	A1	0518
50069.9	W4HHK	Collierville, TN	EM55MM		A1	070
50070	N1LF	Maylene, AL	EM63	1	A1	1119
50070	KL7KY	Wasilla, AK	BP51	75	A1	0619
50070	W7WKR	East Wenatchee, WA			A1	0218

Freq	Call	Town	Loc	Pwr	Mode	Last Rept
50070	WA7X	Fairview, UT	DM49HO		A1	0919
50070.2	NR0X	Martelle, IA	EN42HB	10	A1	1119
50071	W3DOG	Ocean City, MD	FM28EI	45	A1	0919
50071	KM4YDM	Eustis, FL			A1	0419
50072	PY2XW	Campinas, SP	GG67LF		A1	Op?
50072	KK9H	Northfield, IL	EN62CC	8	A1	0120
50072	KF5KOI	Blanco, TX	EM00TA	10	A1	1219
50072	W5HN	TX	EM13SJ		A1	0719
50072.2	KD4YDD	Decala, GA	EM84AB	10	A1	1219
50072.5	EA8SIX	Canary Is	IL28GD	10	A1	0518
50072.5	N7DGI	WY	DN62		A1	0818
50072	N7SCQ				A1	0320
50072.5	W6NIF	Fresno, CA	DM06MM		A1	0617
50072.5	K3BFD	Bedford, PA	FN00		A1	0617
50072.9	N4VBV	Sumter, SC	EM93TW	5	A1	1219
50073	VO1FU	Goulds, NL	GN37OL		A1	0819
50073	K0KP	Fredenberg, MN	EN36VW	100	A1	0718
50073.4	EA3SIX		JN11		A1	1018
50074	W5RP	San Angelo, TX	DM91		A1	0220
50074	W0FY	Chesterfield, MO	EM48RP		A1	0619
50074	NZ5F		EM56JB		A1	1019
50074.5	SV3BSF	Petra	KM08UA		A1	0120
50075	KM4HAM	Pompano Beach, FL	EL96WG	20	A1	0618
50075	WA5SLO	Lucedale, MS	EM50RW	30	A1	0819
50075	VR2SIX	Hong Kong			A1	0819
50075	W7PFR	Mineral, WA	CN86	12	A1	0519
50075	W2UTH	Bloomfield, NY	FN12HV	1.5	A1	0120
50075	KA7BGR	Central Point,OR	CN82	40	A1	0919
50075	NL7XM	Easton, PA	FN20IP	3	A1	0220
50075	AA1OV	Benson, AZ	DM41TW	10	A1	0919
50075	LW2ETU	Avellaneda	GF05TH			0220
50075.5	WB9OTX	Versailles, IN	EM79IB	10	A1	0120
50076.3	NF8M	Southfield, MI	EN82IL	5	A1	0120
50076	K6CS	Redding, CA	CN80TP		A1	0718
50076	W4IT	Sassafras Mtn, SC	EM85OB	1	A1?	0108
50076	CS5BLA		IM57PX			0220
50076.3	YV5LIX	Caracas	FK60CA	5	A1	0719
50076.5	WR9L	Bradley, IL	EN61BD	5	A1	1219
50077	K7AZ	Nr Lompoc, CA	CM94SO	0.5	A1	0119
50077	VA2TKH	Sept-Iles, QC	FO60TE		A1	0819
50077.6	N0LL	Smith Center, KS	EM09OW	60	A1	0120
50078	N2GHR	Nr Centereach, NY	FN30LU	8	A1	1119
50078	WA4FC	Carson, VA	FM17FE	2	A1	0220
50078	KC5SQD	Missouri, TX				0919
50078.7	K1EB		DM43MM		A1	0119

Freq	Call	Town	Loc	Pwr	Mode	Last Rept
50079	JX7SIX	Jan Mayen	IQ50RX	10	A1	QSY
50079.0	K3CX	Philadelphia, PA	FM29JW			0120
50078.9	W8IF	Nr Selma, OH	EM89CR	100	A1	0120
50079.4	TI2NA	Volcan Irazu	EJ89BX	2	A1	0520
50080	XE1H	Guadalajara	DL80HQ	25		0519
50080	LU6ENC					0120
50080	K6FRC	Tracy, CA	CM97HP	100	A1	0819
50080	ZS1SIX	Cape Town	JF96HB		FSK	0419
50080.1	FK8SIX	Noumea	RG37FR	15	F1	1219
50080	4X4SIX	Jerusalem	KM71NU	5	A1	0120
50080.4	VO1STA		GO21		A1	0517
50100	YO8RIX	Nr Dorchoi	KN37FW	50mW	A1	0718
50081.3	PT9BCN		GG29RN	10	A1	0120
50082.5	LU4FW	Pueblo Esther	FF96RW	8	A1	0320
50083.5	CX1AA	Montevideo	GF15LC	3	A1	0220
50085	PP1CZ	Vitoria, ES	GG99UQ	5	A1	0120
50085.5	LU7YS	Tierra del Fuego	FE49IU		A1	1110
50125	VE7CUP					0220
50276	KM4HAM	Pompano Beach, FL	EL96WG	10	JT65	0919
50281	VK4RHT	Atherton	QH22RR	10		0919
50282.1	VK4RTL	Townsville	QH30JP	3	A1	0919
50284	VK4RSC	Nambour	QG63LF	7		0419
50285	VK4RTT	Bunya Mountains	QG63TC	100	A1/JT65B	0719
50288	VK2RHV	Mt Sugarloaf, NSW	QF57SC	25	F1	0619
50289	VK2RSY	Nr Sydney	QF56MH	25	A1	0619
50293.7	VK3RMV	Mount Dundas	QF02WH	15	A1/FSK	NonOp
50295.1	VK3RMH	Wattle Glen, VIC	QF22OH	10	A1	QRT
50295.0	VA2NQ	Mount Yamaska, QC	FN35NL	5	A1/PI4	0617
50297.9	VK7RST	Mt Nelson, Hobart	QE37PB	15	F1	0419
50304	VK6RSX	Nr Dampier, WA	OG89II	50	A1	1119
50306	VK6RBU	Bunbury, WA	OF76WR	20	A1	0219
50310	VK8VF	Darwin, NT	PH57	10	A1	1117
50313	VK6JR	Dunsborough, WA	OF75LY			0217
50315	VK5RBV	Barossa Valley	PF95MK	12	A1/JT65A	0219
50320	VK5VF	Mt Lofty	PF95IA	8	A1	1119
50321.1	ZS5SIX	Hilton				0419
50342.7	VK4ABP	Longreach, QLD	QG26			1217
50350	I0JX	Rome	JN61GW			0917
50400	IW3FZQ	Monse	JN55VF	10	A1	0120
50401	IQ8KK	Cosenza	JM89BK	10	A1	0320
50402	OY6BEC	Faroe Islands	IP61NV			1119
50405	SR6VHF	Gora Chelmiec	JO80CS			0119
50406	F5ZHQ	Hyeres	JN33AC	20		0677
50406	SV2JAO		KN10CL	10		0320
50407.5	IH9/IZ0ANE	Pantelleria I	JM66AS	5	A1	0918

Freq	Call	Town	Loc	Pwr	Mode	Last Rept
50409	ED1YBR		IN52JW			0619
50410.3	SR2KDS		JO94II			0819
50412	OH1SIX		KP11			QSYhr
50413	ZB2SIX	Gibraltar	IM76HD	5	A1	0719
50413	C30SIX	SantJuliadeLoria	JN02SK	10	A1	0819
50414	OK0SIX	Ostrava	JN99CT	5	A1	1019
50415	ED8YAJ	La Palma	IL18CP	10	A1	0220
50415	SR5TDM		KO01KX			0120
50415	SV5SIX	Mt Ilias	KM36XG	3	A1	1019
50415.8	IW1DTK		JN61VG			
50416	EA4KM		IN80BE	50		0218
50417	OH9SIX					QSYhr
50418.5	F1ZFE	Sarreguemines	JN39OC	8	A1	1018
50419	IZ1EPM	27km NE Turin	JN35WD	12	A1	0819
50420	IQ4FE	Fidenza	JN54AS	6	A1	0819
50421	HB9F	Nr Bern	JN37RA			0320
50422	F5ZMT	Nr Guingamp	IN88GN	10	F1	0619
50422	S55ZRS	Nr Dobovec	JN76MC	8	A1	1218
50424	E73SIX		JN92ER		A1	0718
50425	PI7SIX	Rotterdam	JO21FV	50	A1	1219
50425	9H1SIX	Attard, Malta	JM75FV	7	A1	0619
50426	SR9FHA	Choragwica	KN09AX	4	A1	1119
50427	IW0DTK	Minturno, LI	JN61VG		A1	0618
50428	OH2SIX	Lohja	KP20DH	80	A1	0120
50429.8	IZ3OHR	Verona	JN55LJ		A1	0320
50430.3	HG7BVA	Gyomro	JN97QK	5		0819
50430	IW8PNY	JM89CX				0120
50432	F5ZKY	Mont des Alouettes	IN96LV	5	A1	0819
50433	OH7SIX	Tuupovaara	KP52JH			
50436	D4C	Monteverde	HK76MV	30	A1	0818
50437	ES0SIX	Hiiuma Island	KO18KX	15	A1	0819
50438	OE5XHE	Nr Linz 1120m asl	JN78DN	10	A1	0120
50440	ON0SIX	Waterloo	JO20EP	5	A1	0220
50443	GB3MCB	St Austell	IO70OJ		A1/PI4	0220
50445	JW5SIX	Hopen Island	KQ26MM	10	A1	0819
50445.5	ED1YCA	Gijon	IN73AL	5		0818
50446.9	JW7SIX	Kappe Linne	JQ68TB		A1	0819
50448.4	FX4SIX	Neuville	JN06CQ	5	A1	0818
50448.7	IQ0AM		JM49OF		A1	1119
50449	I4JEE		JM54TU	3	A1	0220
50449	JW9SIX	Bear Island	JQ94LM	10	A1	0817
50450	SR1DAR	Darlowo	JO84EK			0614
50450	SV3BSF					0320
50451.0	LA7SIX	Bardu	JP99EC	30	A1	1019
50453	PA3FTL	Nibbixwoud	JO22MQ	50		1218

Freq	Call	Town	Loc	Pwr	Mode	Last Rept
50453.8	HG8BVD	Kisrata	KN06HT	10		0618
50455	LA8SIX		JO59FB	25	A1	0320
50456	VK5RBV	Barossa Valley	PF95MK	12	JT65/A1	1119
50457	TF1VHF	Altfanes	HP84WL	25	PI4,A1	0819
50457	IW0DAQ	Genzano di Roma	JN61HQ	2	A1	1119
50459	IQ4AD	Serione Parma	JN54DT	8	A1	0719
50460	SA1CKI	Visby	JO97DP	5	A1	0220
50461	GB3NGI	Ballycastle	IO65VB	21dBW		0220
50462	SR5FHW	Warszawa	KO02KH	3	A1	Planned
50464	GB3LER	Lerwck	IP90ID			0619
50465	GB3IOJ					?
50466	OE3XAC	Kaiserkogel	JN78SB		A1	0819
50466	OZ4BMDS	JO75KB			CW/MG	0119
50468	SK3SIX	Ostersund	JP73HC	7	A1	1219
50471	OZ7IGY	Toelloese	JO55WM	25	PI4/A1	0120
50474	IQ5MS	JN54AA				0120
50475	OK0NCC	Kukrak	JN79EW	2	F1A/F1D	0919
50474.6	ED7YAD	Malaga	IM76QO	15	A1	1019
50476.1	CS5BLA	Aldeia de Chaos	IM57PX	2.5	A1	1019
50476.7	YM4SIX	KM56IV				1019
50477	IZ3GWJ	Nr Lendinara	JN55TC	5	A1	0819
50479	JX7SIX					QSY hr
50480	JH8ZND	Chitose	QN02UW	10	A1	0718
50480.1	HG1BTV		JN86MN			0717
50481	F1ZFB	Ste Lheurine	IN95TM	10	A1	1219
50483	DF0ANN	Moritzberg Hill	JN59PL	2	A1	0819
50483.0	DB0DUB	Nr Gangelt	JO31AA	1	A1	0519
50483	DB0HGW	Greifswald	JO64QC	3	A1	0719
50485	OH5SIX		KP30	10	A1	QSY hr
50489	OH0SIX	Stalsby	JP90XI	3	A1	0220
50490	F6IKY	Louhans	JN26OP	3	A1	0819
50490	JG1ZGW	Tokyo	PM95VP	10	A1	0519
50493	SL2ZZU	Lulea	KP15CO	1	A1	0917
50499	IQ0A					0318
52342.6	VK4ABP	Longreach	QG26DN	10	FSK	0712
52433	VK3VFO	Morwell	QF31FS	10		1217
52438	VK3FGN	Mildura	QF15CT	3	A1	NonOp
52490	ZL2SIX	Blenheim	RE68		F1	NonOp

Note: This beacon list is based on one compiled by Martin Harrison, G3USF, for the Propagation Studies Committee of the Radio Society of Great Britain (see www.keele.ac.uk/depts/por/50.htm). We are grateful to G3USF for permission to use the list. Please send corrections and updates to m.harrison@keele.ac.uk (Last updated March 2020.)

APPENDIX G – LOCATORS

The IARU Locator System, usually just called 'Locator', provides a means of pinpointing stations throughout the world. It is most often used by operators on 50MHz and above as a means of calculating the distance between two stations (for use by operators on the upper microwave bands, it can have eight digits, though only the first six are dealt with here). The system is based upon latitude and longitude.

There are three sizes of 'rectangle'. The largest, known as a 'field', is 20° of longitude (east-west) by 10° latitude (north-south), and is designated by two letters. Most of Britain is in IO field, as shown below.

The next rectangle, known as a 'square' (though it is actually neither truly square nor rectangular!) is 2° of longitude by 1° of latitude. One hundred squares make up one field and these are given numbers 00 in the south-west corner to 99 in the north-east, e.g. Dublin is in IO63. Finally, each square is divided into 576 'sub-squares', 5 minutes of longitude by 2.5 minutes of latitude, and given letters from AA to XX.

To find out your locator, first use a map of your area to determine your exact latitude and longitude, then use the map and the diagrams opposite to pinpoint your locator. Computer programs and on-line calculators are available to do this more easily: on-line calculators are at www.levinecentral.com/ham/grid_square.php; https://k7fry.com/grid; www.pwcontest.org.uk/NGR2Loc.html and www.mapability.com/ei8ic/maps/gridloc.php

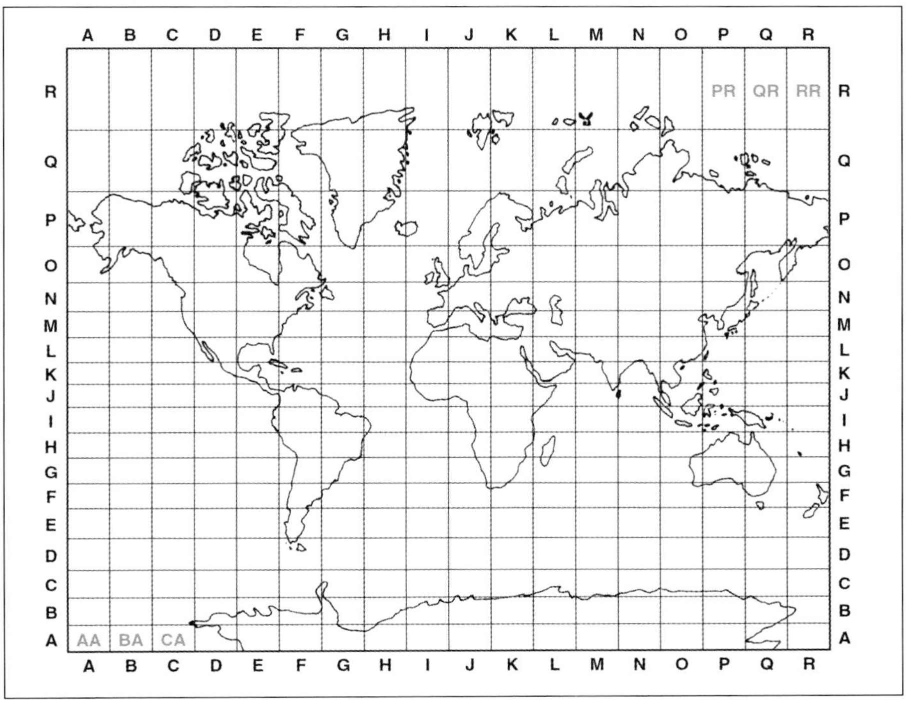

The IARU Locator system may be used throughout the world without repeats. The map above shows the fields that make up the first two letters of the Locator. Examples are shown at two of the corners.

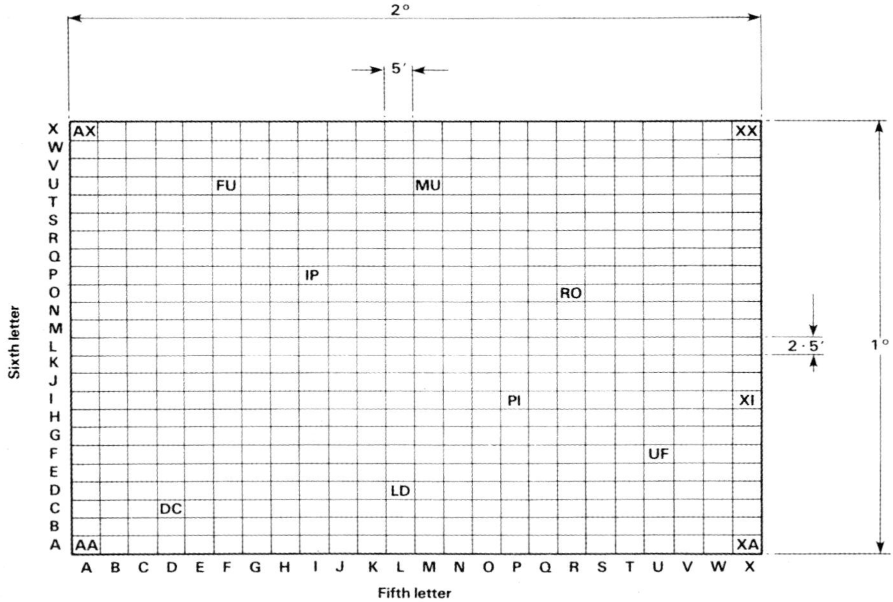

A square (the numbered part of the Locator) is divided into 576 sub-squares, designated AA to XX. Each sub-square is 5' W–E and 2.5' N–S.

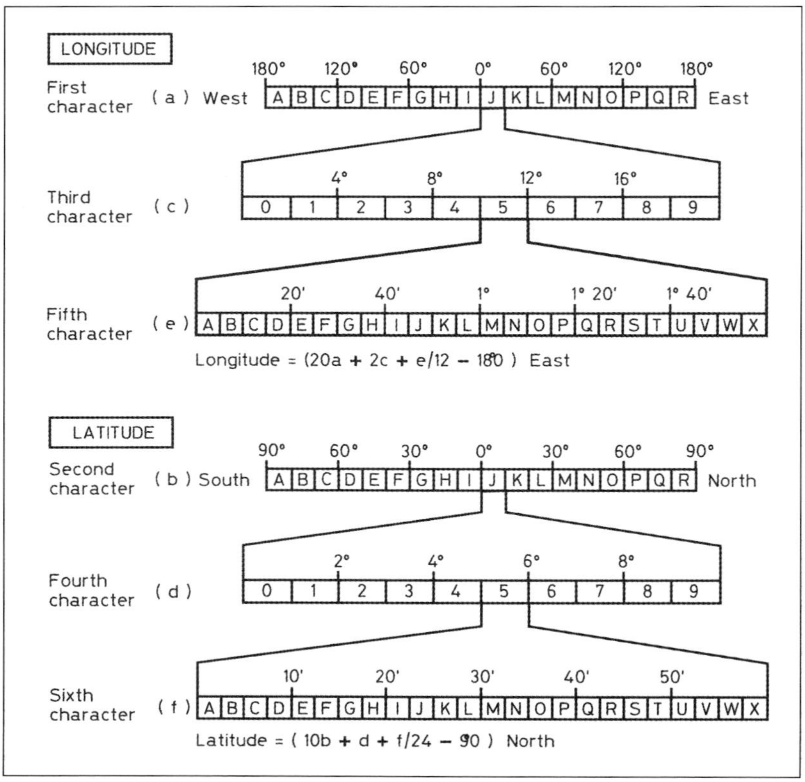

The final two letters of the Locator may be calculated thus.

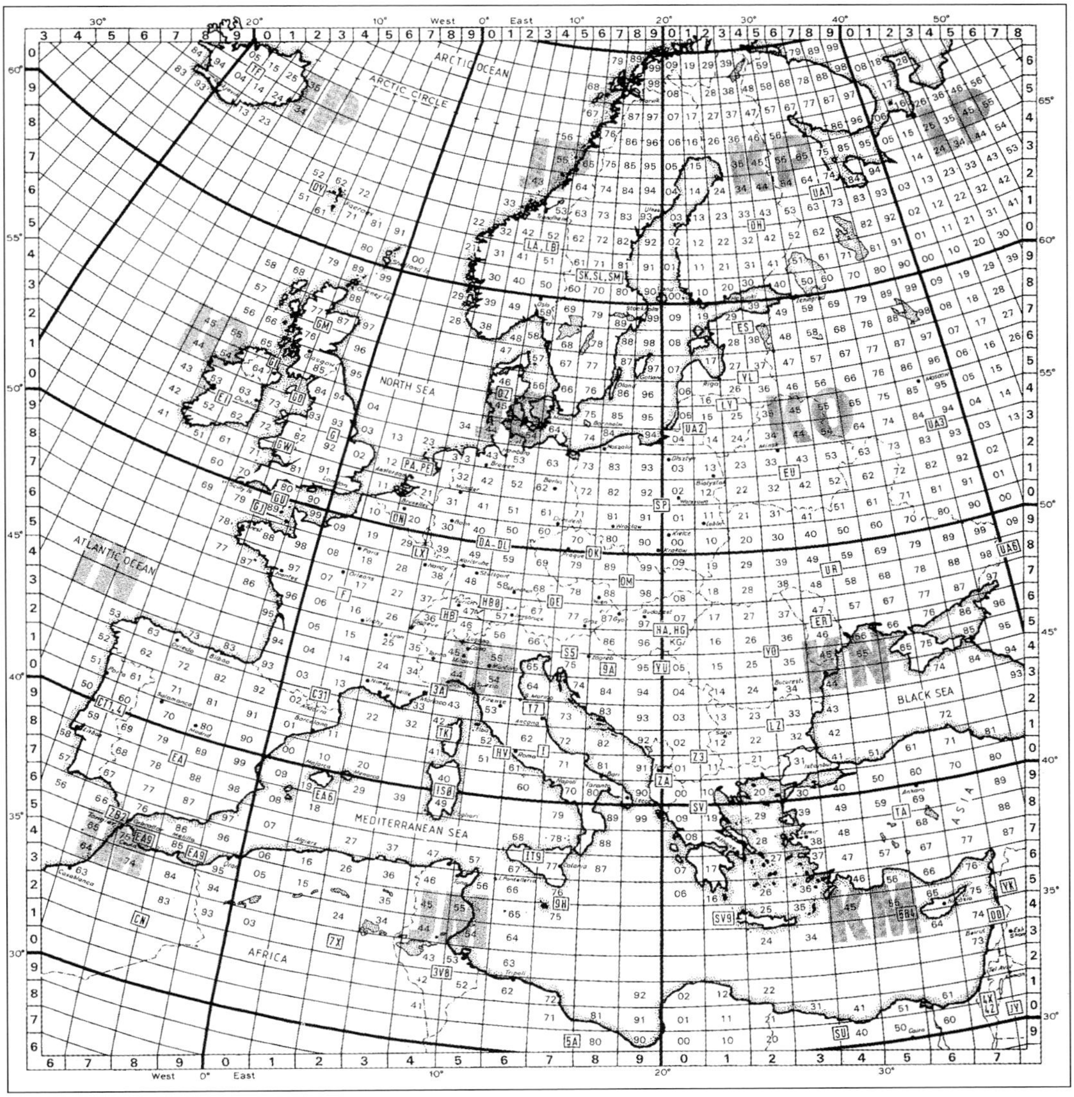

IARU Locator map of Europe.

APPENDIX H – RSGB 6M BAND PLAN

MHz	Necessary Bandwidth	UK Usage
50.000-50.100	500 Hz	Telegraphy Only (except for Beacon Project) Note 2 50.000-50.030 MHz reserved for future Synchronised Beacon Project (Note 2) Region-1: 50.000-50.010; Region-2: 50.010-50.020; Region-3: 50.020-50.030 50.050 MHz Future International Centre of Activity 50.090 MHz Intercontinental DX Centre of Activity (Note 1)
50.100-50.200	2.7 kHz	SSB/Telegraphy - International Preferred 50.100-50.130 MHz Intercontinental DX Telegraphy & SSB (Note 1) 50.110 MHz Intercontinental DX Centre of Activity 50.130-50.200 MHz General International Telegraphy & SSB 50.150 MHz International Centre of Activity
50.200-50.300	2.7 kHz	SSB/Telegraphy - General Usage 50.285 MHz Crossband Centre of Activity
50.300-50.400	2.7 kHz	MGM/Narrowband/Telegraphy 50.305 MHz PSK Centre of Activity 50.310-50.320 MHz EME 50.320-50.380 MHz MS
50.400-50.500		Propagation Beacons Only
50.500-52.000	12.5 kHz	All Modes. 50.510 MHz SSTV (AFSK) 50.520 MHz Internet voice gateway (10 kHz channels), (IARU common channel) 50.530 MHz Internet voice gateway (10 kHz channels), (IARU common channel) 50.540 MHz Internet voice gateway (10 kHz channels), (IARU common channel) 50.550 MHz Image/Fax working frequency 50.620-50.750 MHz Digital communications 50.630 MHz Digital Voice (DV) calling 50.710-50.890 MHz FM/DV Repeater Outputs (10 kHz channel spacing)

MHz	Necessary Bandwidth	UK Usage
50.500-52.000	12.5 kHz	All Modes. 51.210-51.390 MHz FM/DV Repeater Inputs (10 kHz channel spacing) (Note 4) 51.410-51.590 MHz FM/DV Simplex (Note 3) (Note 4) 51.510 MHz FM calling frequency 51.530 MHz GB2RS news broadcast and slow morse 51.650 & 51.750 MHz See Note 5 (25kHz aligned) 51.770 & 51.790 MHz See Note 5 51.810-51.990 MHz FM/DV Repeater Outputs (IARU aligned channels)

Note 1: Only to be used between stations in different continents (not for intra-European QSOs).
Note 2: 50.0-50.1MHz is currently shared with Propagation Beacons. These are due to be migrated by Aug-2014 to 50.4-50.5 MHz, to create more space for Telegraphy and a new Synchronised Beacon Project
Note 3: 20 kHz channel spacing. Channel Centre frequencies start at 51.430 MHz.
Note 4: Embedded data traffic is allowed with digital voice (DV)
Note 5: May be used for Emergency Communications and Community Events
Note 6: Digital Experiments to support innovation may occur at 50.6, 51.0 or 51.7 MHz with a 100kHz maximum bandwidth

LICENCE NOTES:
Amateur Service 50.0-51.0 MHz – Primary User
Amateur Service 51.0-52.0 MHz – Secondary User: 100W (20dBW) max, Available on the basis on non-interference to other services (inside or outside the UK).

Notes to the Band Plan
ITU-R Recommendation SM.328 (extract)
Necessary bandwidth: For a given class of emission, the width of the frequency band which is just sufficient to ensure the transmission of information at the rate and with the quality required under specified conditions.

The use of Amplitude Modulation (AM) is acceptable in the all modes segments but users are asked to consider adjacent channel activity when selecting operating frequencies.

Foundation and Intermediate Licence holders are advised to check their licences for the permitted power limits and conditions applicable to their class of licence.

APPENDIX I – ARRL 6M BAND PLAN

6 Metres (50 – 54MHz)

50.0-50.1	CW, beacons
50.060-50.080	beacon subband
50.1-50.3	SSB, CW
50.10-50.125	DX window
50.125	SSB calling
50.3-50.6	All modes
50.6-50.8	Nonvoice communications
50.62	Digital (packet) calling
50.8-51.0	Radio remote control (20-kHz channels)
51.0-51.1	Pacific DX window
51.12-51.48	Repeater inputs (19 channels)
51.12-51.18	Digital repeater inputs
51.5-51.6	Simplex (six channels)
51.62-51.98	Repeater outputs (19 channels)
51.62-51.68	Digital repeater outputs
52.0-52.48	Repeater inputs (except as noted; 23 channels)
52.02, 52.04	FM simplex
52.2	TEST PAIR (input)
52.5-52.98	Repeater output (except as noted; 23 channels)
52.525	Primary FM simplex
52.54	Secondary FM simplex
52.7	TEST PAIR (output)
53.0-53.48	Repeater inputs (except as noted; 19 channels)
53.0	Remote base FM simplex
53.02	Simplex
53.1, 53.2, 53.3, 53.4	Radio remote control
53.5-53.98	Repeater outputs (except as noted; 19 channels)
53.5, 53.6, 53.7, 53.8	Radio remote control
53.52, 53.9	Simplex

APPENDIX J – RSGB 4M BAND PLAN

MHz	Necessary Bandwidth	UK Usage (Note 1)	
70.000-70.090	1 kHz	Propagation Beacons only	
70.090-70.100	1 kHz	Personal Beacons	
70.100-70.250	2.7 kHz	Narrow Band modes	
		70.185 MHz	Cross-band activity centre
		70.200 MHz	CW/SSB calling
		70.250 MHz	MS calling
70.250-70.294	12 kHz	All Modes	
		70.260 MHz	AM/FM calling
		70.270 MHz	MGM centre of activity
70.294-70.500	12 kHz	All modes channelised operations using 12.5 kHz spacing.	
		70.3000 MHz	
		70.3125 MHz	Digital modes
		70.3250 MHz	DX Cluster
		70.3375 MHz	Digital modes
		70.3500 MHz	Internet voice gateway (Note 2)
		70.3625 MHz	Internet voice gateway
		70.3750 MHz	See Note 2
		70.3875 MHz	Internet voice gateway
		70.4000 MHz	See Note 2
		70.4125 MHz	Internet voice gateway
		70.4250 MHz	FM simplex used by GB2RS news broadcast
		70.4375 MHz	Digital modes (special projects)
		70.4500 MHz	FM calling
		70.4625 MHz	Digital modes
		70.4750 MHz	
		70.4875 MHz	Digital modes

Note 1: Usage by operators in other countries may be influenced by restrictions in their national allocations.
Note 2: May be used for Emergency Communications and Community Events.

LICENCE NOTES: Amateur Service 70.0 – 70.5 MHz – Secondary User: 160W (22dBW) maximum. Available on the basis of non-interference to other services (inside or outside the UK).

Notes to the Band Plan: (see Appendix H)

Index